HEAL THE WORLD

Heal The World

DAVID ICKE

Gateway

Gateway
an imprint of
Gill & Macmillan Ltd
Hume Avenue, Park West
Dublin 12
with associated companies throughout the world
www.gillmacmillan.ie

1 85860 005 7

Cover design by Studio B of Bristol
Text illustrations by Jackie Morris
Text set in Bembo 10.5 on 13 by
MetraDisc of Castleton, Rochdale
Printed by ColourBooks Ltd, Dublin

Diagram on p.14 is from "The Chakras" by
Naomi Ozaniec, published by Element Books

The paper used in this book comes from the wood pulp of managed
forests. For every tree felled, at least one tree is planted,
thereby renewing natural resources.

A catalogue record is available for this book
from the British Library.

5 7 9 8 6

Contents

OTHER BOOKS BY DAVID ICKE

It doesn't Have to be Like This (GreenPrint)
Truth Vibrations (Gateway)
Love Changes Everything (Thorsons)
In the Light of Experience: *autobiography* (Warner)
Days of Decision (Carpenter): *also available as a cassette* (Gateway)

I was told through a psychic that I would write five books in three years, which would present the spiritual truths to the world. It seemed ridiculous because at the time I knew nothing about the subject. But here I present to you the fifth book I have written in three years. Shortly before I started *Heal the World* another psychic told me that my next book (this one) would be the most important I have written so far. I believe that to be so.

To Lyn Gladwyn (Yeva),
a source of love and support and
one of the few people I know with a
sense of humour as daft as mine!

1

Who is Healing Whom?

Heal the world. Three little words, with enormous implications.

In fact they tell only part of the story. *Heal yourself, heal the world* would be more accurate; and *heal yourself, heal the world and let the world heal you*, would be even more so.

For nothing happens in isolation. There are no islands set apart. I am you and you are me. We are all part of One Consciousness, One Mind, that is Creation. It is the amalgamation of all thought, experience, understanding and wisdom we know as God or, as I prefer, the Infinite Mind.

You are God. I am God. We are all God. So when we say *heal the world*, it is not a one-way process. It is not a case of one person or group of people appearing over the horizon like some spiritual cavalry to save the planet. Yes, we can help to heal this glorious expression of life called Planet Earth, but the Earth can heal us, also. We are part of her and she is part of us. We are all the same consciousness at different levels of evolution and understanding.

If we don't heal our own wounds we cannot heal the planet. *Physician, heal thyself.*

2

Love Yourself, Love the World

I have seen, in the light of extreme experience, how people constantly project out into the world what they think of themselves. I have observed outwardly aggressive people venting their inner hatred of themselves on others around them. Their victims are merely mirrors for them to thrash out in anger and frustration at their dislike, even contempt, for all that they are and have been. This is the real motivator behind so much violence, crime, and negative behaviour of all kinds.

People were persuaded by the media at one stage that I was mentally ill because of the information I was presenting. As a result, many people came along to ridicule and laugh when I spoke at universities and public meetings. They did not realise that in doing so they were making a mighty statement about themselves, not me. How sobering it would have been had they asked themselves what imbalance must exist within for them to have gleaned pleasure and enjoyment from ridiculing someone they believed to be mad. When you do not respect others you are merely saying you have no respect for yourself.

How often have you noticed how people get angry and upset with those who act in ways that they themselves also act? How often have we said of such people: 'It's okay when they do so and so, but when someone else does it they go crazy'? What we are seeing is that person reacting against someone who mirrors back at them the character traits they dislike in themselves. On a couple of occasions when people have laughed at me in the street, I and my family have stopped and laughed back at them, mirroring their behaviour, to see what happened. The reaction was extraordinary. They became very angry, even outraged, as they came face to face with their own behaviour. When we see others expressing what we don't like about ourselves we feel very uncomfortable, and that can manifest in outward anger and condemnation. We appear to be angry with another, but really we are angry with ourselves.

Humanity's problem in general, if not in every 'individual', is that far from respecting itself, liking itself, loving itself, it hates itself. Humanity is awash with guilt, fear, and lack of self-esteem, and this goes back through scores of previous physical lifetimes as well as this one. It is this guilt, fear, and self-hatred that is expressed through division, war, conflict on all levels, and in the belief that somehow material 'success' can compensate for the spiritual and emotional black hole that grows like a cancer within so many.

And by spiritual I do not mean religious. The two are constantly confused. Spiritual for me means a reconnection with the One Consciousness, an understanding of our part in the eternal scheme of things and our infinite potential for love and creation. Religion has hijacked spirituality and largely abused its name to build empires of myth and power, engineered and perpetuated by the manipulation of fear and guilt.

Spirituality is the answer while, in stark contrast, religion is part of the problem.

I talk of the need for self-respect and self-love, and yet religion constantly tells us we have sinned and must feel guilty about all that we are. I heard a lady brought up in a Roman Catholic convent say how she was forced to recite every morning that she was unworthy. The whole religious, economic and social system is designed to make us feel either guilty, fearful, unworthy, dissatisfied or envious – indeed often all of them, and more, at the same time. This is terribly destructive both for people and the planet.

Self-love and self-respect are crucial to the future well-being of the earth because the extreme negative energies we have created and continue to create by the way we feel about ourselves, and therefore others, have taken this planet to the brink of non-existence. We are going to reverse this process – and to do that we need to understand the structure within which we all live and evolve. We need to appreciate how we are affecting the balance of the earth and each other with every thought.

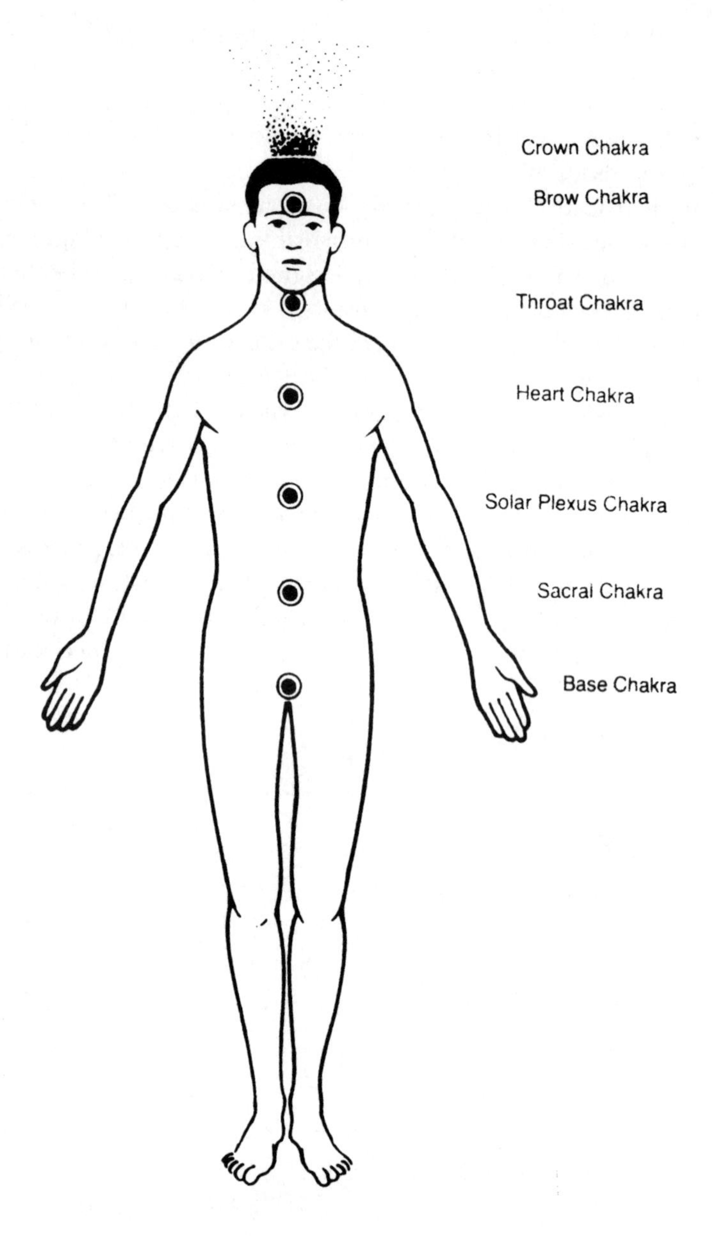

The Seven Chakras

3

The Real Us

The physical body is a vehicle for the mind, our consciousness, to experience this physical level. The real thinking, feeling, eternal us is not the body, but a series of energy-fields working as one. I call this the mind to avoid the religious connections associated with the word 'soul'.

Joining these fields together are seven main vortex points known as *chakras*. These are at the top of the head (crown chakra), the forehead (brow chakra), the throat (throat chakra), the centre of the chest (heart chakra), the solar plexus (solar plexus chakra), the navel (sacral chakra) and in the area of the sexual organs (base chakra). There are others, but these are the major ones.

Each chakra has a different role to play. The solar plexus chakra, for instance, works with the emotional level, and this is why we feel our emotions in that area. 'I've got butterflies in my stomach' – as many people say to describe nervous moments. The heart chakra is the point of balance in the chakra system and also the point from which the energy called love is expressed.

Your mind, everything that is the real you, has worked through endless physical bodies in endless situations over incomprehensible amounts of time. What you are experiencing now is simply the latest of them, although potentially the most important. I see the physical body as a genetic spacesuit. Just as we need a spacesuit to exist on the moon, so the mind, which is an amalgamation of non-physical energy-fields, needs a physical vehicle, the body, through which to experience this physical world.

The colour or design of your spacesuit is irrelevant, except as another form of experience. Those who condemn black people should know that they have almost certainly been a black person themselves in a past life, or they will be in the future to balance out their eternal experience. The divisions of colour, sex, ethnic background, countries and religion, which cause so much pain and conflict, are the manifestations of human misunderstanding. No matter what the design or colour of the genetic spacesuit, we are all still of the One Consciousness. Any apparent division is an illusion in the collective mind of the human race.

Our consciousness has re-incarnated on its journey of experience into physical bodies that were black, white, red, yellow, male, female, and into life situations that made us rich, poor, Protestant, Catholic, Muslim, etc. The fact is that someone born in the United States is not more special than someone born in Mexico; nor is someone born in England more special than someone born in Peru; nor someone who is white than someone who is black. Go down to the seashore, hold a droplet of water in your hand, and tell me how that droplet can possibly be any more or less special than the billions of droplets you see before you in the ocean.

Yet look what this misunderstanding of life has done to this world. The physical body is perceived to be the real us and so people are judged on the colour and origin of the ancestral line of that physical body. It is like judging an astronaut on the design of his spacesuit. Healing ourselves and the planet will be so much easier once we realise that.

Creation consists of an infinite number of levels of evolution or 'frequencies'. The frequencies of all the television and radio stations broadcasting to your area exist in the same space you are standing or sitting in now, and, in the same way, so do the frequencies, wavelengths, of creation. When you tune your radio to one of the broadcasting frequencies around you, that is the only one you receive. To the radio at that time, that is the only wavelength there is. It is the same with the mind.

At any point in our evolution we tune ourselves to one of creation's wavelengths, and that becomes our reality. At this moment we are tuned to this dense physical wavelength, and this appears to be everything that is. But at the instant we call physical 'death', the mind leaves this wavelength and moves to another, because its period of experience on this level is over. Many millions of people have had what are called near-death experiences, when they 'die' and are then revived.

They describe how their consciousness left the physical body. That is all that happens at what is known as 'death'. The fear of death terrifies so many, when really it doesn't exist. We don't die – we are reborn onto another wavelength. Everything on that wavelength looks just as 'solid' as it does here, and all those other 'solid' worlds exist in the same space as this one.

How can this be? For the same reason that the broadcasting signals can pass through the walls of your home to reach your radio. The wavelength of the wall is so different from that of the broadcast signal that one passes through the other as if the other isn't there. When people see 'ghosts', they often appear as a misty figure. This is because the wavelength the 'ghost' is working on is different to this one. When your radio dial is not quite on the right station you get a fuzzy, weak, less-than-sharp reception, with maybe bits of other stations coming in as well. In visual terms that is what is happening when we see 'ghosts', which are merely minds not in a physical shell. They appear misty and transparent because we are not on the same wavelength as they – so our reception is similarly fuzzy, weak and less-than-sharp. But if we were on the exactly same wavelength they would look as 'solid' as we do.

I keep putting the word solid in quotes because, in truth, although we use that word all the time, nothing is actually solid. Everything is energy vibrating at different speeds. At slower speeds the energy is denser and so can appear to be 'solid', but it is not. Everything that exists in any form is vibrating energy. We are vibrating energy, too. The speed our mind-energy is vibrating denotes our level of evolution. The higher our 'vibrations', the higher our state of evolution, knowledge, understanding and wisdom.

We also project energy out into the world with every thought, because each time we think we create an energy-field, positive, negative, or a balance of the two. This is the basis of telepathy. One mind creates a thought energy-field or thought-form and another decodes that field and turns it into words. You will notice that many times two people will have the same thought at the same moment. One reason for this is that one person thinks of something, creates a thought-field, and their friend unknowingly decodes that thought-field, while believing it is their own thought. Hence you have the reaction of 'Well I never. That's just what I was thinking!'

The question has long been asked: *What is energy?* Energy is movement, motion, and this motion is created by thought. Vibrating energy particles are not particles in motion, but particles of motion. The particles are created by thought-patterns, and when those patterns change or disap-

pear, the particles those patterns have created also change or disappear. So when you think love you create and generate the energy of love. The patterns formed by thoughts of love create the energy called love, which can transform the world. Think negative and you surround yourself with negative energy which, you, yourself, have created. This is how thought creates and also how we create our own reality. When I speak of negative and positive energies I am referring to negative and positive thought-patterns. Thought and energy are the same.

Another key to understanding how we affect the Earth and each other with every thought is to appreciate that there needs to be some kind of balance between the negative and positive energies within us and around us, if we are to avoid extremes. If we generate too much positive energy, we become 'ungrounded'. We float off in some imbalanced dream-like state and lose touch with the practical side of life. If we generate too much negative energy it shows itself in anger, conflict, a desire for power and control, and other unpleasant behaviour. And what are the most powerful creators of negative energy? Guilt, fear, self-hatred and lack of self-worth, which, as I have outlined, lead to outward expressions of anger, violence, conflict, and the need to dominate and control. This is where we came in.

Over vast amounts of what we call *time*, humanity's self-hatred, encouraged by forces that wish to destroy the planet, has generated fantastic amounts of negative thought-fields, which in turn have caused some appalling events. These have led to an even lower sense of collective self-worth which has then led to even more negative thought energy... On and on it goes. It is a spiral that must be broken if humanity is going to survive the process now in motion of keeping the planet in incarnation. This tidal wave of negative energy has imbalanced the energies all around us, and the mind of the planet, the Earth Spirit. All the negative energy humanity has generated she has had to absorb, because we all live within her gigantic energy-field, her mind.

What happens when we become very emotionally upset? We stop thinking straight. Think of the implications, then, if you consider that it is the consciousness we call the Earth Spirit, Mother Nature, *Gaia*, whatever name you prefer, that keeps this planet together on every level. It is she who balances the weather patterns, the geological structure of the earth, the regeneration of nature, and the mixture of oxygen in the atmosphere that allows life as we know it to exist at all. If the mind that is responsible for this becomes imbalanced and ceases to think straight, because of all that humans do and have done to her, just imagine what the consequences could be for everyone living on Planet Earth. I will not seek to hide it from you. Those are the implications that face us now, today, if we do not heal ourselves, so the Earth Spirit can be healed and cease to be bombarded with negative energy.

What I seek to do here is pass on what I have learned from my own experiences, and offer some answers to the question so many are asking today: *What can we do?*

4

Think Freedom

Freedom of thought is the greatest of all freedoms. From this all other freedoms come. It is also the very foundation of self-healing and self-respect.

The system, the status quo that currently controls the physical world, can only survive and continue on its ultimately suicidal course if it persuades enough people not to think beyond the limitations it wishes to impose for its own protection. Once you free yourself from the conditioning we undergo from the media, the education system, and other sources of 'information', you see this human-created world for what it is. Ridiculous. Thinking for yourself is fatal for the system and must be avoided at all costs if it is going to maintain its increasingly tenuous grip on the human mind.

This is why the system seeks to create structures designed to self-perpetuate the status quo. You only pass your exams and progress within the system if you answer the questions set by the system, in a way that is acceptable to the system. There was a time when if you had said in an exam that the earth was round and not flat you would have failed. The same applies today with countless other views held to be 'fact' by the status quo. You only become a doctor if you say in your exams what conventional medicine wants you to say. So the way the body is affected by our non-physical levels, past life experiences, the law of karma, and negative-positive imbalances, goes on being ignored by mainstream medicine, to the enormous detriment of the patient.

Examinations are just one way in which the system self-perpetuates and passes on its nonsenses across the generations. There are countless others. Look at the world around you and you'll see so much that is ludicrous which only survives because it was here when we were born into the world. It survives because not enough people have opened their eyes and asked the question why? Why do we do things this way? Who decided it? On what basis? And is it right that it should continue? Apathy, taking the line of least resistance, is the attitude of mind the status quo depends upon for its survival.

I would offer the following credibility test for something that was already here when we were born: If the same thing was being done today for the first time what would our reaction be? Take the Church as an example.

Imagine that the Church did not exist and along came a group of people with a book of ancient texts. "We don't know who wrote them" they would say "And we don't know when they were written or in what circumstances or how much they have changed over the centuries for political reasons or how much of the original meaning has been lost in the translation. But we know it's the word of God and all accurate even though it contradicts itself constantly and in parts is a justification of violence, slavery, and the suppression of women."

Now imagine what we would think if the BBC came out

and said that they were going to give free, guaranteed airtime to this new religion, and that this airtime would not be there to debate and question, but to give this religion the platform to say whatever it liked to promote its beliefs. Imagine also our reaction if the British Government announced that it was going to make compulsory the teaching of this new religion in the schools and that it would be taught on the basis that it is historically accurate and not just a story that a certain number of people believe to be true.

Think how we would react. We would be on the streets with placards protesting against this imposition by an authoritarian state. We would be shouting "Moonies", "thought control", "indoctrination", "mad cult", and the like. But wait. What I have described is precisely what happens in Britain and many other countries all over the world today. Where are the people on the streets protesting? Where are the shouts of "mind control"? Nowhere. Why? Because all this was here when we arrived and so we meekly accept it.

When I was speaking in Houston, Texas, I went into a delicatessen for a cheese sandwich. When you don't eat meat in Texas you survive on cheese sandwiches most of the time. When the waitress came across I told her what I would like. Her face dropped immediately.

"I can't serve you a cheese sandwich here" she said, "If I did this place would be shut down."

"What? Shut down because you serve me a cheese sandwich?"

"Yes".

I was informed that I was in a Jewish delicatessen and it was against the rules to serve cheese in a place where meat was also served. (I should point out that this deli was in the corner of a supermarket where you could buy all the cheese and meat you liked.)

"Who decided that you can't serve cheese?" I asked, by now utterly stunned at what I was hearing.

"I dunno, the Jewish hierarchy I guess."

"And when did they decide that?"

"I dunno, thousands of years ago I think."

"So because of what the Jewish hierarchy decided thousands of years ago, I can't have a cheese sandwich in Houston in 1993?"

"That's about the size of it, I guess."

The other interesting aspect to this story is that the more I asked questions, the more uncomfortable and indeed angry the waitress became. This was because she, too, knew that the situation was ridiculous – but look at the implications for her, a Jewish woman, of standing up and saying so. Her business would have been closed down by the Jewish leaders and she would probably have been castigated by her family and friends. Much easier to take the line of least resistance, allow the inherited nonsense to go on unchallenged and pass it on to the next generation intact.

Also while I was in Houston I was told another story which sums up what I feel is the vital point I am making here. A woman explained how she always used to cut the corners off a ham before she put it in the pan until one day her husband asked why she did that.

"I don't know," she replied. "My mother always used to do it and so I do it."

"Why did your mother do it?"

"I don't know, she just did it. What does it matter?"

"Call your mother and ask her why she did it."

She made the call.

"Mom, you know when I was a child, why did you always cut the corners off the ham before you put it in the oven?"

"Because my pan was never big enough."

Apply that simple story across the great tapestry of life on Earth today and you will see how much destruction and imprisonment of thought continues only because we don't have the vision to ask "why". If we are going to break the cycles of perpetuated illusion and misunderstanding, it is essential that we ask that question constantly, and that means challenging situations and ways of thinking that do not wish to be challenged. That is going to take courage because courage is essential to world healing.

The most important way the status quo survives is through fear. Freedom of thought is really freedom from fear. The two are indivisible. The system has been so effective in selling nonsense as wisdom that we now have a situation in which the most credible explanation of creation is immediately ridiculed or condemned by reflex action. Talk of reincarnation, the eternal nature of the human consciousness and other wavelengths of life, has been largely laughed at by the system's propaganda machine, the media.

But promote what is, in my view, the most ludicrous idea that this teeming haven of interdependent life emerged by accident – and this is seen as perfectly acceptable. This makes so many frightened to say what they really believe because they think it will not be acceptable to their friends, the neighbours, the people they work with or the media. This fear suppresses the knowledge which enormous numbers of people believe is the true basis of life.

I have had letters from people saying they have long believed in all the themes I am highlighting, but they dare not even tell their wives or husbands because they fear the reaction. There was an ambulanceman I met who said that he had realised that medical science had great limitations of understanding when he began to feel a tremendous healing energy working through him when he was with a patient. But he dared not tell his colleagues because he felt they would say he was crazy. The system has been very clever and highly successful in suppressing the information that is likely to bring it down. This it has done by the skilled use and manipulation of fear. Emotions like fear and guilt have their part to play in the overall life experience, but on earth it has got out of hand. Those emotions are no longer experiences, but monsters that control the world.

So the first stage of self-healing is the release of fear. What is there to fear? Fear of what our friends may think, our families, or the neighbours and people at work? They have every right to think what they like – but then so do you. If they cannot respect your right to think something different from them, then they are making a statement about themselves, not you. And just ponder for a moment. You may see those whose reaction you fear as people with minds that are closed. But wait. If you suppress your feelings and views because you fear what people around you with closed minds will say, you are allowing your life to be controlled by closed minds.

Being different is the only way to be in our society as it is today. An interviewer said to me once: "You are very controversial aren't you?". I replied that if I was not seen as controversial I would be very worried. What would it say about me if what I was saying was considered acceptable to a system that is destroying the world and causing so much pain, misery and violence in the process?

Yes, I'm controversial – and I'm proud of it. But most people don't live their lives in the way they wish or think is right for them. They live them in a manner they think acceptable – non-controversial – to others. The message the system finds unacceptable is:

THERE IS NOTHING TO FEAR.

The whole 'power' of the media was turned on me and my family and will continue to be so for a time yet. But here we are, stronger than ever, more confident and self-assured than ever, and with more true friends all over the world than ever before.

Step out of your closet of fear, lift your head high, and speak your truth to the world. Then freedom will be yours.

5

Prisons of the Mind

Ironically, even some organisations set up apparently to challenge the status quo are also fearful to go too far. In other words, far enough. You need to beware of how their limitations can be transferred to you.

The environmental movement has played an important role in highlighting what we are doing to the planet, in our desire for more, more, more, and the constant demand for an expansion of production and consumption – economic growth. But there is every sign as I write that environmental organisations who base their analysis on conventional science have stopped evolving their understanding of the wider picture within which they operate. Much of the reason for this is that they have allowed themselves to become mesmerised by human 'science', and fearful of challenging that 'science'.

They see 'science', much of which is not scientific at all, as the crutch to their credibility. Humanity has become so conditioned to believe that scientists are the fountain of knowledge (not true) that environmentalists feel that if they can support their argument scientifically it will increase their credibility in the public mind. That may well be the case, given the conditioning I have just described, but the real question they could be asking is: *are the scientists correct in what they say?*

On the one side the environmental movement is opposing the use and production of endless potions and poisons created by conventional science, and yet at the same time it looks to that same conventional science for the answers to the problems thus created. Fear is the reason. The result is an imprisonment of thought. Green pressure groups will not let go of conventional scientific analysis because they fear they will lose credibility with the public, who have been conditioned to worship the scientist. This is not to say that all that system-serving science says is not valid. What it lacks, I would suggest, is a vision that understands that its knowledge and research involves only a tiny fraction of all that is. Without this understanding it must by definition be offering the world a highly imbalanced view. As Socrates said: 'Wisdom is knowing how little we know.'

The green movement is at a fork in the road. It must choose between fear – which will lead it to oblivion – or freedom of thought – which will lead it to the next stage in its journey of evolution. That same choice is now being offered to every man, woman and child on this planet. While the Greens have been portrayed as unconventional, their biggest weakness, I feel, is that they have not been unconventional enough. Again we come back to fear.

The answers to environmental decline will not come by persuading politicians and industrialists that urgent and fundamental changes are required. Industrialists are under pressure from others to increase production and profits because that is what the system insists of them. Collectively politicians do and say what they think is necessary to be elected.

You have a situation in which the propaganda of the economic, political and social system encourages people to see life in a certain way, and those people demand of their politicians the policies they believe will give them what the system has told them are the indicators of 'success' – money, status and possessions. The politicians react to those demands, and the destruction of the planet not only goes on, but gains momentum with every year, in the suicidal pursuit of more money and possessions for the minority at the expense of the majority and the Earth.

I am not attacking politicians or industrialists here. They are trapped within a system that insists they act in a certain way, otherwise they don't become politicians and industrialists in the first place. It is not people I am challenging, but the thought-patterns that control those people. No one in a physical body runs this system. What I describe as *The System* is really thought-patterns which dominate the minds of billions of people. The only means to change what politicians do is to change what they need to do and stand for to get elected. That means changing the collective thought-patterns of humanity, so we think in another way and demand something very different from those who represent us.

You can have all the legislation you like on environmental protection, but unless people start to think on a higher level of understanding, there will always be ways around it. And, let's not be naive here, legislation will constantly be too little too late, because pressures on politicians not to allow environmental concerns to impinge on their obsession with economic growth will forever delay or dilute what is really needed. As then-President Bush said before the Earth Summit in Brazil in 1992: "I will agree to nothing that harms America's economic growth".

So, yes, it is highly desirable for the environmental movement to encourage recycling, non-polluting technology and lifestyles that are in harmony with the Earth, and not at war with her. Everything that lifts some of the burden from Mother Earth is to be welcomed, and all this is most certainly part of the process of healing the earth – or, to be more accurate, slowing down the rate at which we assassinate her. But it is but a speck of sand in a desert unless it is supported by a spiritual revolution and a reconnection of humanity to the One Consciousness, the Infinite Mind of all that is.

Only then will we have access to a level of understanding and perception in which harming the earth would not even occur to us, let alone require legislation to discourage it. Only then will we cease to generate the destructive energies and vibrations that are continually imbalancing the Earth Spirit, and reducing her capacity to repair, regenerate and hold together the natural flow and balance of life on her physical form. Time to move on, Green Movement of the World! Time to see a much bigger canvas. Time to let the walls around your thinking fall away, and to allow the fear of going beyond the conventional scientific view to disperse and emerge as a new and greater vision.

What has happened to at least large sections of the Green Movement is a warning to everyone. There is no end to the journey of understanding. It is truly eternal. There are only stages along the way. The problem with movements and 'isms' is that they can become ends in themselves. An organisation or movement is set up at a particular stage in the beliefs and perceptions of those involved. But instead of saying "This is what we believe now, and we will evolve that view should new information come to light", those organisations from that point onwards see new information as a threat if it challenges the original beliefs on which their organisation was formed. For this reason they stand still, and the agenda moves beyond them. In the end they become irrelevant and they crumble.

Look at the Church. You would think that life ceased to evolve, certainly spiritually, two thousand years ago. It's okay to stand in a pulpit and talk about this angel speaking to so and so, or God speaking to someone else. But if you say you are communicating with other levels of consciousness today, you are branded as evil and working with the devil!! The Church's desperation to hold on to a set of beliefs, decided largely by a Roman Emperor and a group of bishops in 325AD, has led to more wars, deaths, torture, violence and mind-control through fear and guilt than anything else in human history.

These 'spiritual' empires have become a vehicle for personal and collective power which depends for its survival on holding back the tide of open thinking, which would otherwise expose the nonsense and contradictions of the view it has sought to impose upon the world. The Church, like so many organisations, reflects in its desperation to survive by the suppression of knowledge, the system as a whole. Indeed the Church is a fundamental part of the system. The King or Queen of the United Kingdom is crowned by the Archbishop and declared a 'defender of the faith' or, it would be more appropriate to say, a symbolic defender of the status quo for Church and State.

'Isms' are prisons of the mind, be they Roman Catholicism, Communism, Capitalism, Socialism, Spiritualism, or whatever. Once you are involved in an 'ism' you are, in subtle and less-than-subtle ways, encouraged not to allow your thinking to expand beyond the walls of the 'ism'. If you expand beyond these walls, you are seen as a threat, a traitor, not a 'true' Catholic, Socialist, Communist, or whatever. This can persuade you to suppress what your mind and heart are saying, and so you stay true to the 'ism', and not true to yourself. It is vital in the process of healing ourselves, and therefore the world, to be very careful of organisations that stand still and see new ideas and information as a danger rather than the next wonderful stage of evolution and understanding. That means most of them.

Humanity is taught to see evolving views as a weakness, when it is the ultimate strength. Evolving your understanding is seen as admitting you were 'wrong', or that you made 'mistakes'. Poppycock. There are no mistakes. What we call *mistakes* are essential experiences that open up our understanding. The two perceptions of the same experience are very different. Mistake = guilt and lack of self-worth (negative). Learning experience = new understanding, evolution, and greater appreciation of potential and self-worth (positive).

Be true to yourself and what your heart is telling you. That is what really matters. Of course work with others if you feel that is best for you – and I will talk more of this as we proceed – but always follow your own feelings and intuition of what is right and wrong. Have the confidence to stand firm with what you believe, and don't allow others to apply pressure that will impose their often rigid and unevolving beliefs upon you. If you do, you will not heal yourself – and certainly not heal the planet.

'Isms' and organisations force people to compromise their views so the group can show a 'united' front. This is considered essential, for some reason. I smile when I hear that a political party is 'disunited', and therefore (presumably) in trouble. What is so wrong about having a group of people who do not agree on everything, and are prepared to say so? Surely that is highly desirable, unless you want a world full of clones – which, of course, the system does. 'Isms' are terrified of being seen as anything but united, and so you get people either saying what they do not believe to be right, or suppressing their real views in an effort to protect some facade of unison. The revolution of humanity now underway will not be diluted by such fear. It is a revolution of self, and compromising what we think and say to suit the desire for a unity of view has no part in this. Saying what we really think and respecting another's right to think differently *is* the revolution.

I feel that large parts of the environmental movement are

threatening to become just another 'ism', another imprisoned belief-system. It is up to everyone inside the movement to stop that happening or, failing that, to break free now and see wider possibilities and other ways of healing the planet. Even areas within that vast tapestry of thinking known as the New Age movement are falling into the same trap. They are showing serious signs of becoming just another religion, with their teachers and gurus and 'living gods on earth'.

Some parts of the New Age movement are becoming another industry emerging to exploit the gathering public interest in these subjects and the confusion of those turning their minds to these areas for the first time. There are so many 'services' on offer and so much past/present life 'counselling' available which is simply nonsense and irrelevant to your personal growth and transformation. Often the only effect they have is to confuse and divert you, part you from your money, and make you distrustful of the whole subject.

Whenever people want to exploit something, it is essential to complicate it or make it appear complicated. Once they do that, others who do not understand the manufactured complications can be exploited. The economic system of the world does that and so, it should be said, do parts of the New Age movement. The process of awakening and transformation is much more straight forward and simple than some would have you believe. Open your mind and your heart and don't let karmic counsellors, rebirthing tanks, and all the other New Age paraphernalia get in the way. At the very least, the overwhelming majority of it is quite unnecessary.

Beware on your spiritual journey. Listen to people, even dip into organisations here and there if it feels right. But if anyone tells you what you *must* think, or if anyone looks upon you differently because you think differently to them, it's time to go. What can be a good source of information and even inspiration, can also be, if you are not careful, a prison of the mind.

6

Taking Control

The human race, certainly in the 'developed' world, seems to have a desperate desire to pass on to someone else its responsibility to think and act – to an 'ism', church, teacher, cult, rule-book, boss or politician.

We are all encouraged to do this. While governments may talk of the need for self-responsibility they are the last people who would actually want to see it happen. When you take responsibility for your own thoughts and actions, the credibility of the system, governments included, falls before your eyes. It is this ability to get people to pass on their responsibility to the system that has allowed it to maintain its pre-eminence.

One sees this desire to pass on responsibility every day. When you meet an official in a uniform and peaked cap, you often get the answer: "Rules is rules, mate. It's more than my job's worth to go against the rules". The person involved is merely refusing to take responsibility for making a decision because they feel more comfortable surrendering that responsibility to a rule book.

I remember going into a car park once and I went across to the attendant to ask how much it would be for five hours. He quoted a large sum of money.

"But hold on", I said, "In another hour's time it will be 6pm and it says on your board that it is only 50 pence for the rest of the evening after that. Surely I pay the full price until then and the rest of the night should cost me 50 pence?"

"No sir. If you want to qualify for the 50 pence price you will have to come back at six o'clock, take your car out into the street, turn around and come back in."

I waited for him to laugh. He didn't.

"Rules is rules, mate", was his general reply and that made life so much easier for him. No decisions to make, no responsibility to take. The system is slavishly served and the rule book rules, ok?

There are billions who hand over responsibility to those most famous of rule books, the Bible, the Koran, and others. They are desperate to believe that all the decisions they need to make are waiting for them within the covers of books full of contradictions and mis-translations written centuries ago, often by who-knows-who in who-knows-what circumstances.

In the New Age movement there are people giving their responsibility to a teacher or guru, or to a discarnate consciousness communicating through a medium or channeller. By all means use these as sources of information, but in the end we must all make our own decisions on what we feel is right for us. Life on earth is all about self-responsibility, and much as we may try there is no avoiding that if you wish to evolve. The law of cause and effect – karma – sees to that.

The mind has free will to make decisions and learn from the consequences of them. Every day we are facing situations, making decisions (or allowing others to make them for us!), and learning from the result. The law of karma, which I feel is a sort of electromagnetic process, draws to us what we need to balance out our evolution. Everything is energy, even situations and experiences. I feel that through a form of electro-magnetism we are drawing to us the energies – people, situations, etc – that we require. We are constantly experiencing both sides of the balance-point so we eventually see where that balance-point actually is. We only recognise lukewarm because we have experienced hot and cold. In the simplest of terms, what we do to others will be done to us, or, as the Bible says: "What you sow you shall reap". There are many themes of truth within the Bible amid the misunderstandings, but those themes are not unique to that book. The idea that what you sow you shall reap is common to many cultures, beliefs, and civilisations.

Before each new incarnation the mind decides what it wishes to experience in that physical life. What it needs to balance out in its experience or karma will be a fundamental part in that decision. Many of our experiences in this life are to balance out what we have experienced in previous lives. This is not punishment. There is no judgmental God saying you have sinned and you must be punished. We make those decisions, but we make them when we are free from the limitations imposed by the physical body. When we decide a life-plan we do it from a much more enlightened perspective to the one we inherit when we find ourselves looking through our eyes into the physical world. We know that we need the interplay of negative and positive energies to create the understanding that raises our vibratory frequency and allows us to evolve. Without this interplay we would, at best, stand still.

As one psychic communication said: "Balance is a blissful and glorious state, but it is also a static one. To continually evolve to higher levels of understanding requires the interaction of positive and negative energies created by positive and negative experience. It is the interplay of negative and positive energies that allow the vibrations to quicken."

What we do not need, however, are the massive extremes we see on earth. I see evolution as a series of stages between points of negative-positive balance. I feel that when you reach that state of balance you are at the top of one frequency or level of evolution and able to progress to the next one. Once you are there, the negative-positive interplay begins again as you seek the balance-point at that new level of information and understanding. You won't be forever reincarnating into this physical level. When you have reached the balance-point you will move up and onwards from here – indeed the greatest opportunity is being offered for you to do that now in this life.

So what does all this mean in our daily lives? Quite simply that to believe we can pass on our responsibility to someone or something else is an illusion, and in terms of our evolution, a dangerous one. If you follow the instructions of another, instead of making that decision for yourself, it doesn't matter to karma. You will, in karmic terms, need to experience whatever you have made others experience – even if your action was the result of what someone else told you or advised you to do.

For so many reasons, people need to realise that we are all responsible for our every thought and act, and that what we do to others will need to be done to us to balance our evolution. Only in that way will we free ourselves from the rule books, cults, religions, power-structures and other manifestations of mind-control. What others think and do is their responsibility. What we think and do is ours.

7

Not Guilty

Ah yes, guilt. Along with fear it is the most rampant emotional cancer known to the human race. Until we rid ourselves of guilt there will be no self-worth and no healing of ourselves or the Earth.

Why do we feel guilty? Because society, programmed by inherited 'values', bombards us with endless things to feel guilty about. And that's the point, you know. Most of the limitations on what is defined as 'acceptable' human behaviour are mostly not even decided by the generations subjected to those limitations. We inherit them.

I use the example in another book of the slogan I saw on a tee-shirt which said: 'Jesus had long hair'. We don't actually know what the man we call Jesus looked like, but that's by the by. It is taken for granted that this figure had long hair. So why do we have people who go to church to apparently worship this man and then, when they see youngsters with long hair, *they* say: "Look at those long haired lay-abouts –

how can their parents allow them to go out looking like that?"

This reaction is not because long hair is right or wrong. For goodness' sake what does it matter, and what business is it of anyone else anyway? No, the reason long hair is seen by many as unacceptable results from the conditioning those people have allowed themselves to be subjected to. When they were born into this world, they inherited the values of the time which were, in turn, the values passed on from previous generations. So they go through their lives believing short hair is good and long hair is bad. Yet had long hair been acceptable and the norm in our society when they came into the world, they might now be going along the street saying: "Look at those *short*-haired lay-abouts – how can their parents allow them to go out looking like that?"

The values of a society at any one time have little, often nothing, to do with right and wrong. They are only that society's *perception* of what is right and wrong, a perception mostly inherited from the past. There was a time, for example, when the whole idea of slavery was acceptable to our society. But did that make slavery right? Of course not. It was an abomination. Much of what is acceptable in the world would be resisted or ridiculed if it were being started for the first time today. Such 'values' survive only because they were part of society when people were born and they become accepted because of that. But all these so-called 'values', which remain by default, by not being challenged, are the source of enormous guilt which entraps humanity in a web of self-disdain.

That little word *sex* is perhaps the biggest creator of guilt. If you trace back our society's values surrounding sex you will find that their origins lie in religion or the interpretation of religion. The limitations we impose upon ourselves about sex in the last decade of the 20th century were largely decided by religion centuries ago. These are the values that say you should only have sexual intercourse if you have said some words in a church at a marriage service or if you have signed a piece of paper in a registry office. Do those words in a church make any difference to how much love those people have for each other? No. Does signing a piece of paper in the presence of a government official make any difference to that love? No.

Then why on earth is it considered okay to express love physically if you have a marriage certificate, and unacceptable if you don't? Again it has nothing to do with right and wrong and everything to do with society's perception – in this case a perception decided by someone's interpretation of the Bible, the King James version of which, has, according to the enlightened historian Arthur Findlay, at least 36,191 errors in translation. I love my wife Linda with everything I have, but I would not have got married if my awakening had happened over twenty years ago. I do not see it as relevant to the way I think about her. I love her deeply. A piece of paper makes no difference to that.

Going on from society's view of marriage and sex is the belief that it is only acceptable to love one person at any one time. You apparently have to fall out of love with one man or woman and part before you can express love physically to another. Even then to be fully acceptable you have to marry the new partner before any 'hanky-panky' takes place, and you may find that some churchmen will refuse to marry you because they do not believe in divorce.

But let's just look at this for a second. We are all aspects of the same One Consciousness. We are all part of each other. Society is saying to us that we can only express physical love to what is another aspect of ourselves if we have a piece of paper signed on the law's behalf. It is further saying that we can't make love with a second aspect of ourselves unless we ditch the first aspect, somehow cease to love them, and get another piece of paper signed by an official to link us by law to the second aspect. Yes, I know it's absolute nonsense, so why do we allow it to go on unchallenged? Fear, that's why. This is one reason why the elimination of fear from our lives goes alongside the elimination of guilt.

Why should making physical love to other aspects of the One Consciousness affect our relationship and the deep, unbreakable love we may have for the aspect that we may be spending our physical life with? In reality it should have no effect because we are all each other in different genetic spacesuits. But it does have an effect, of course. It causes great pain and emotional upheaval for all involved and can smash relationships into tiny pieces. But what causes that pain and emotional upheaval? Is it that making love with more than one aspect of ourselves is wrong? No, how can it be, when viewed from a wider perspective of life? The pain is caused by what we are all programmed from birth to *believe* is wrong, and what society believes is wrong. The pain is not caused by what happens, but the way we are programmed to *perceive* what happens.

If we were born into a society which understood that we are all aspects of the same whole and viewed each other in that way, then the values we inherit about sex, love and guilt, would be very different. In turn so would our reaction. Not expressing love in a wider sense would be seen as unusual, and the idea of possessing another rather than loving them unconditionally, and seeing physical love with other aspects as a natural part of life, would be the exception rather than the rule. It's merely a matter of the indoctrinated social view of what is 'decent' behaviour and what is not.

Don't get me wrong here. I am not suggesting a sexual free-for-all. And I am not talking about making physical love with every aspect you can find! It may well be that you have only one sexual partner in a physical life because that is what you decided in your life-plan or what you feel is right. The point I am making is that others who have different life-plans and choose other values are not wrong, just as you are not wrong. You are different, that's all, and that is something to celebrate, not scorn. We are born free, we 'die' free, we are eternally free. There are no limitations except for those we invent and impose upon ourselves. It is our mind-controlled society that seeks to turn us into imprisoned robots and, in the case of billions of people, has largely succeeded in doing so.

What I am saying about sex is that we need to follow our intuition. There are many reasons why we may need to make physical love with more than one partner, and if we are honest with ourselves our intuition will tell us what is behind the motivation to do so. I say from experience that if you are meant to make love with someone and you are in close enough synchronisation with your higher mind, then it will happen. The energies that will be switched on at many levels will ensure that it happens. There are scores of reasons why this may be part of a life-plan, particularly at this time:

★It may be for the personal development, awakening and understanding of others around you who have chosen to cope with the society-inspired emotional trauma that their partner's sexual relationship with another creates. Should it all happen very publicly it is almost certainly because the experience is designed to affect the way people in general think, too.

★It could be that energies within both people need to be activated. There is a chakra vortex-point in the area of the sexual organs, and these interact during sexual contact. It can be that these energies stimulate, awaken, each other, and/or they combine to create energies for healing the planet. While misguided humanity sees sex as a physical act, higher levels see it as the bringing together of energy fields, energy-patterns. Particularly if you are working with earth energies it may mean you need sexually to combine energy-patterns with two or even more aspects of the One Consciousness. It

could just as easily mean that you do not. It depends on what you are here to do.

★While society is programmed to see the conception of children as the conception of the whole human being, the higher levels of understanding see conception as the creation of the physical vehicle for the incoming mind. That physical vehicle, the body, needs to have the correct genetic inheritance and energy-combinations provided by the parents for that mind's life-plan to be successfully carried out. If you carry certain genetic codes and energies it may well be that both have to combine with more than one partner to provide the ideal genetic and energy make-up for the different children you conceive. For instance combining with your 'life-partner' may be correct for most of the children you create, but you may need to combine with another energy-pattern to produce the ideal physical and energy combination for another child. Society cries 'adulterer!', while the incoming mind says: 'Thank you very much – this body is just what I wanted!' I have no doubt that the man we call Jesus fathered children. His genetic code and energy patterns would not have been wasted. They would have been passed on.

★It can also be that the two people involved are merely the physical vehicles for energies from higher levels to flow through them to create a physical body. This is where the idea of the 'Immaculate Conception' may have originated.

There is much more to sexual contact than we have understood, and we should not feel guilty about it. The myths and bigotry handed to us from a misguided, Bible-obsessed past mean that we even have two types of children, 'legitimate' and 'illegitimate'. Yet the only difference between the two is a piece of paper saying 'married'! Write it in letters ten feet high: ALL LIFE IS EQUALLY LEGITIMATE.

There are people who frown on one-parent families. They say a child needs two traditional parents, and they seek to impose this in many cases. What a child needs, what all children and all people need, is *love*. That's the key word. I have seen many children brought up in an 'acceptable' family situation who have enjoyed anything but love from their parents. I have also seen children living with only a mother, a father, or a guardian, who are surrounded by tremendous love. If you were a child, which one would you pick if you had to choose? And let us not forget that the incoming mind has control over when and where it incarnates. They know the score before birth, and they make the decision, no one else.

Homosexuality can be a generator of great guilt and pain, because of the way it is viewed by programmed people. But what can possibly be wrong with one aspect of the Infinite Consciousness showing their love physically for another who just happens to have the same design of genetic spacesuit? We are all different, thank goodness. Love is love: and the love of one man for another or one woman for another is no less valid or 'legitimate' than that between a male and female.

The irony is that because of all the guilt, myths and misguided 'values' surrounding the subject, modern society has become obsessed with sex. Without their daily batch of stories linked to sex, what on earth would the tabloid press find to write about? When we begin to appreciate that sex is merely a way of expressing love for another aspect of ourselves and, more to the point, for creation as a whole, then the human race will become less obsessed with sex, not more. The bigger the taboo, the bigger the obsession with it. Fortunately attitudes about sex are changing and that change will become a transformation in the years ahead.

The only thing I would stress again is that it is important to follow your intuition on these things, and beware of illusions leading you into situations with people that are not part of your pre-planned life experience. There have been many life-plans shattered in this way.

But whatever you have done or will do in your life there is no need to feel guilty. Everything you experience is adding to the evolution of everyone else, either because of what others experience as a result, or from what you have experienced and pass on to the Infinite Mind as a whole. There are no such things as mistakes, only learning from experience, and the only true love is unconditional love, which allows each other to learn from experience without grudge or judgement. Other versions are not love in its purest sense.

That's another thing about guilt: the way we feel bad about hurting those we love. This is quite understandable, but what is true love? To quote a channelled message I have used in every book: "True love does not always give the receiver what it would like to receive, but it will always give that which is best for it".

If you have acted towards someone in a negative way it may have been your free will overriding your higher mind. But it could just as easily have been your higher mind offering a learning experience to the victim of your negative behaviour. As I have already said, we need a combination of negative and positive experiences (energies) for the negative-positive interplay to take place, that quickens our vibrations and allows us to evolve through a frequency. Negative experiences are as essential as positive. It is a balance of the two we are after. The same applies to the way people react to us. In my autobiography, *In the Light of Experience*, I have highlighted how those who have acted towards me in the most negative fashion have done the most to awaken my inner understanding.

Sometimes we need a negative experience to help us think and awaken to life and ourselves as we really are. I have known people who have changed their entire thinking and become much happier and more contented when very negative events like cancer, a heart attack or business crash, have forced upon them a need to reassess their lives, perceptions and priorities. I have no doubt that you can look back at your life and feel really guilty about giving someone a bad experience, but there is an excellent chance that your behaviour was pre-arranged to help that mind evolve. If it was your free will and not pre-arranged, well you have learned from the guilt you feel. Now is the time to let it go. Whatever we do, there will always be the chance to put right what we have done, if not in this physical life, then in another, or on another level of reality. Life is forever.

The other guilt-creator I see scattered around like confetti is the idea that every negative experience we have must be 'our karma'. Some people feel really guilty because they think that a terrible event in their lives must be a sort of karmic retribution for something they have done in the past. Well, yes, perhaps it might be a balancing of experiences. But it is not retribution or punishment, it is the balancing of energies and understanding – a gift which they have chosen to give themselves to speed their evolution. It is also a gift they are giving to everyone else.

How often have we avoided unpleasant situations because we have seen others get into trouble? We have learned from another's unpleasant experience. Instead of condemning such people for their behaviour, we could thank them for their contribution to enlightenment. And they themselves need to let go of their guilt and understand what they are doing for others. As we hear so often on the news bulletins, as people in great distress say: "I only hope that some good comes of this, and it never happens to anyone else". When one person's suffering alerts humanity to the need for change, it can prevent so many others suffering in the same way, and worse, in the future.

And, let me emphasise that not every negative event in our lives is our 'karma'. We might be being subjected to another's free will, and there are many on the planet at this time who are here to undergo experiences, often highly negative, to work out some of humanity's collective karma. Karma works on many levels. Besides personal karma, there is family karma, national karma, planetary karma, and more. These are the energy-imbalances created by these various groups over the centuries, and minds have come into incarnation to help clear these imbalances. This can mean going through some awful experiences to act as sort of spiritual kidney machines, attracting negative energies from the sea of energy all around us, and transmuting them through their energy-fields into positive or balanced energy. You tune into certain collective thought-patterns and 'live' them. This can appear very negative and painful at the time. Then through that experience you change your thinking and break that collective thought-pattern. Some people describe this as like hoovering the frequency of collective negative imbalances. Thought-patterns can be like railway lines which lead you to think and behave in a certain way. Pattern-breakers travel along those lines, but then change their whole way of thinking. This helps to break that pattern, that vibratory state, and in turn others are less affected by it until it no longer exists at all. It is at this level of thought-patterns that the status quo is being overturned, day after day.

You could quite easily be one of those pattern-breakers, and many of your experiences may not be the karma from your own past behaviour, but the collective behaviour of a group, or humanity as a whole. We don't know for sure, and the only way we can react is to accept what comes our way, and deal with it in the most positive, loving manner we can – loving especially ourselves, and not feeling guilty.

We feel guilty when we cannot cope with a situation and it is seen as a sign of weakness to 'go under' emotionally for a while. Big boys are told not to cry, and they feel guilty if they do. This is all programmed bunkum. If a few big boys allowed themselves to cry more often, that emotional release would stop many a heart attack. There are times when things become too much to handle, and our emotions collapse. This is natural, and nothing to feel guilty about.

As you begin to move ever faster along your spiritual path, be kind to yourself. One form of guilt is feeling guilty about being less than pure and perfect with every thought and act. Let's be honest, some people, whether through free will or pre-arranged experience, are a pain in the backside sometimes. We all are when we want to be. If we feel guilty about every thought that confirms that some people can irritate us, we are going to feel guilty quite a lot. Let those thoughts take their course and then let them go. Later you will find it within yourself to see that those who irritate us are on a journey of experience, too, and coping with that very irritation helps us to learn and evolve. As time goes on, we find that we don't get so affected by another's negative behaviour, and eventually we are not affected at all.

It is the same with the process of loving unconditionally. It is just that: a process. Guilt can come if you are frustrated with yourself at not being able to see everyone in those terms immediately. It will come, but let it come in its own time without your feelings of guilt and inadequacy getting in the way. Love yourself, be kind to yourself, and you will then be ready to love and be kind to others.

Forgive yourself when you fall below the ideals you seek, and you will find that you can forgive others when they act in ways that you don't like. Forgiveness of others can only come after forgiveness of self.

8

A Little Respect

There are some who believe that without fear and guilt society would collapse into chaos and anarchy. It is a perception that encapsulates the system's perverse idea that you can only keep humanity in check by the imposition of negative emotions. The opposite is the case.

I was once questioning the concepts of fear and guilt in a debate at the Oxford Union when a student stood up and said that it was only fear and guilt, especially fear, that held society together. Without fear of retribution such as imprisonment, he contended, people would do even more awful things to each other than they do now. That student was articulating the thoughts of billions of people around the world. I worked in a television department where it was thought that the way to keep the staff working hard and conscientiously was to keep them in a constant state of uncertainty and concern for their future. It is the same in all areas of life. How many children are controlled by the fear of parental violence?

But surely if you build and seek to control a society through those most powerful of negative emotions, you will create a negative society. It will be a society in which the negative will dominate thinking and belief-systems. A society, for example, that will concentrate on what we can't do or are not allowed to do, rather than emphasising the limitless potential we all have.

The idea that the controlling emotions of fear and guilt are irreplaceable is a gigantic myth. They can be replaced and will be replaced in the years and decades before us by the positive and self-balancing mechanism we call *respect*. That's the key word of this time and the future. Respect for ourselves and respect for everything. That is a positive foundation on which to build a society, and on which to build our own personal outlook on life.

I was in a radio studio a little while ago, waiting to be interviewed. All morning, apparently, they had been having a phone-in on the subject of fox-hunting, and now I was listening to an interview with a wonderful man who had gone out from England to Bosnia, in an effort to raise the money to rebuild a bombed orphanage. Three times in that chat, the interviewer made the point that now they were talking about an issue that really mattered, after spending so much of the morning talking about fox-hunting. He clearly thought that the horrors of fox-hunting were not as important as the horrors of Bosnia. But what is the difference except in degree? The same state of mind, the same lack of respect for self and others, is behind both situations.

Once you believe you can create a league-table of life-forms on the basis of which are more important and worthy of respect than others, you are entering very dangerous ground. What is the difference, except where you draw your line of expendability, between saying: "I can justify killing foxes for pleasure because they are lesser forms of life", and saying, "I am a Bosnian Serb and so I can justify killing a Bosnian Muslim, because they are lesser forms of life"? If we treat animals badly, we'll treat people badly. The same thought-patterns are responsible for both.

So often I find that people don't see they are behaving in the same way as those they oppose. There is no difference between the violence and pain a scientist forces animals to suffer in a vivisection laboratory, and the violence against the scientist by a tiny minority of extreme animal liberation activists. Violence is violence and begets more violence. It is the same thought-pattern manifesting. When I said this at a rally opposing vivisection, the vast majority of the audience enthusiastically agreed because they knew this was so. But a handful of people afterwards told me that one day they would hang me, because if I was against violence on vivisectors I could not possibly care about animals!

The word is *respect*. If we respect the rights of all life, if we see them all in the terms that we see ourselves, and if we relate what happens to them with how we would feel if it happened to us, we will begin to revolutionise the values which pervade this planet. Respect for all life, no matter what form it may take, will mean that the fox-hunter will cease to kill the fox and the Bosnian will cease to kill other Bosnians who don't share his culture, religion and ethnic inheritance. The elimination of fear and guilt both within us and without will be replaced by respect.

Once you have respect for the Earth on a global scale (positive), you don't need legislation to say that you must not pollute beyond what the law allows (negative). If you respect all people of whatever colour or creed as equal expressions of the One Consciousness (positive), you no longer require legislation to say that you must not discriminate against someone on the grounds of their colour or creed (negative).

Fear and guilt create wars and conflict. Respect for self, and so the world, will make war impossible.

But for this to happen, respect has to be a two-way process. Take the case of the people we in Britain call 'new age travellers', who wish to have a nomadic lifestyle going from place to place, living in their buses and caravans. They ask others who don't live like that to respect their right to have a different lifestyle. Quite right, too. What arrogance to believe that, because society says it's 'normal' to live in a permanent dwelling, others who choose otherwise are lesser human beings. What is missing is respect. Behind that abuse and contempt for travellers is our old friend fear – the fear of something that is different to the status quo, which so many see as their 'security'. So I am one hundred percent behind the travellers' right, and anyone else's right, to live in the way they feel is good for them.

But. It is clear, too, that there are elements – and only elements – within these travelling groups who, while demanding respect for their way of life, do not show a similar respect for others. They may see people who complain about their behaviour as toffee-nosed, middle-class system-servers, but whatever they are, they have a right to live their lives as they feel fit, too. They have a right not to be disturbed by loud music when certain travellers stop in their area, and they have a right not to see that area covered in rubbish. If respect were the motivator behind both lifestyles, the conflict would disappear. If conventional lifestyles respected the rights of others to live differently, and vice versa, travellers and non-travellers could live in peace and harmony. Until that happens, the problems will go on.

You are a sacred and special aspect of Creation, and every other aspect is equally sacred and special. Respect yourself for what you are, and you will respect others for what they are.

Now that's a world I want to live in.

9

Judge and Jury

Now we come to the most popular sport and pastime known to the human race: judging others. Under starters orders... and they're off!

"I think it's absolutely disgusting that she's got another man – and so soon after her husband died, too."

"Have you seen what our neighbour looks like? What a tart!"

"I think all people who run businesses are scum – all bosses exploit the workers."

"Do you know that we now have some blacks in our street. It's outrageous – they are no better than monkeys in the trees."

That last judgment was actually said in my hearing. It's amazing what we say about each other. But the system encourages us to judge – that is part of the way it controls. It goes even further and sets out how people should be judged, not even by what they *are*, but by what they *own*; not by how they care, but by what they wear; not by the size of their heart, but the size of their house. Possessions and income are the system's measurements of success because it needs us to buy that delusion if it is to persuade us to buy all the other possessions, services and trinkets that it has to sell to survive.

And yet real success in a physical lifetime is not the accumulation of money and possessions, it is to find inner peace, contentment and happiness, and to complete experiences designed to speed our evolution. That may mean having a lifetime with a small income and few or no possessions. Judging another's success by what they own is the height of misunderstanding, I would suggest.

But what is the judgment of others in any form when you take it back to the level of self? It is an excuse, a diversion, which stops us looking at our own behaviour. While we are judging others and pointing out all the things we feel are wrong about them, we can ignore those things about ourselves that might deserve a little examination. It is also a way of protecting our own belief-systems. While we judge others constantly from the perspective of our belief system, we can avoid questioning and evolving what we believe. In other words, when we judge others, we are in danger of standing still. One of the main motivations behind judging others is the fear of looking at ourselves with a view to change.

Humanity has become conditioned to view change with fear and trepidation. The status quo does not want change except within strict limitations, because significant change means, by definition, the end of the status quo. Given a choice, turkeys would not vote for Christmas or Thanksgiving. It's the same with the personal and collective status quo. Our security has ceased to come from within ourselves – instead it has become outwardly expressed. Security has become a 'decent' home, a 'good' job, an unchanging world around us, and the prospect of all three continuing into the future. We fear outward change because we fear change within ourselves. We resist it, and one way we do that is to take our static 'values' and judge everyone and everything by

those 'standards'.

Yet when we judge others, we are giving them permission to judge us – and would your life stand up to scrutiny even if that judgment were based on the same criteria and 'values' that you apply to people when you judge them?

As I have said already, another big reason for judging others harshly is because they are reflecting back at us traits of our own behaviour, and that can really make us explode. Somehow we think that if we oppose those behaviour traits in others, we can ignore that we ourselves have them, too. Only by accepting that can we take steps to change. But it is worth stressing that being honest about ourselves, and seeing areas that we wish to develop and change is not the same as judgment, just as questioning what happens in the world is not judgment of people. Looking at ourselves honestly, but kindly, with understanding and compassion, speeds evolution. Looking harshly and judgmentally at ourselves can hold back evolution. Judgment of ourselves is not the way. But constantly questioning the belief systems that control our behaviour is very necessary.

I know there are a few people who see it as quite wrong for me to take information into the public arena and challenge the system of life that we have. They see it as being judgmental. But what are they doing, by that same criterion, except judging me? We could go around in circles here. By making information available I am not saying that people have to believe it, only that they have a right to hear it along with the views of everyone else. To deny them that right, would be to leave them to the imbalanced flow of information given to them by the status quo, which filters out anything that seriously questions its own existence. What I am doing is not judging people, but challenging thought-patterns by showing that alternative ways of thinking and explaining life do exist, contrary to what the status quo wishes us to believe. Judgement of people and the questioning of belief and behaviour systems are not the same. Without the questioning of thought-patterns, there is no evolution, either for ourselves or the whole.

What makes the judgment of other people, as people, most ridiculous is that we have no idea why they are acting in the ways they are. It may be their free will and, if so, they will learn from it. But it could just as easily be an experience pre-arranged to change our own view and awaken us to other potential possibilities and ways of thinking. We could well be making harsh judgments about people who love us so much on higher levels that they are giving us an experience designed to help our own evolution.

Let us put an end to judgment of ourselves and others. Then, and only then, will loving unconditionally become possible.

10

The Great Awakening

All that I have said has been designed to highlight how we are controlled by what comes in through the eyes and the ears, and how we can walk free from that control. From here on, I am looking at where people can go from that point of understanding. The most common question I am asked is: "Okay, I accept the basis of what you are saying. What can I do now?" I will offer some answers in the rest of the book, but only you can decide if they feel right to you.

When our minds incarnate in each physical life, not all of the mind is subject to the limitations of the physical body. Only some levels of the mind work directly through the brain. The higher levels of our consciousness, the subconscious, super-conscious, etc, guide what I call the 'physical consciousness' through a series of life-experiences. It would be of no value if that part of our minds experiencing this physical level – the 'physical consciousness' – knew what it had come to experience and how it hoped to react. There would be no learning.

The physical consciousness (lower self) experiences, and our higher levels (higher self) guides us through those experiences. At the end of each physical life all becomes at one again, and all the experience and knowledge is absorbed by the whole.

When we feel drawn to certain people and places, or when our intuition urges us to take a certain course of action, this is our higher self speaking to us in thought-energy being passed down from higher to lower levels. When we say 'what a small world' and 'what a coincidence', they are not coincidences at all. They are the work of higher selves making things happen so we can experience them with certain other minds. This might be for karmic reasons, or because a task needs to be done for the benefit of creation. How we react to those situations, using our free will, decides the next stage of our lives. If we make a decision that the higher self considers inappropriate for our evolution we will consistently find ourselves facing the same situation again and again, until we make another choice. From that point the situation ceases to recur, and another stage opens up.

How often do we go against our intuition and regret it? Over enormous amounts of time, humanity has been sending out disruptive, negative thought-patterns (energies), as the programming has controlled our thinking. This has imbalanced the energies within which we all live, and encouraged us to think and generate even more negative energy. One major consequence of this has been that the communication channels between higher and lower selves during an incarnation have been much less powerful amid this ocean of increasingly negative-dominated energy on this frequency. What comes in through the eyes and the ears – the system – has overpowered the messages coming down from higher levels of ourselves. In that one sentence you have the biggest reason why the planet is in the state it is.

One of the ways we have lost touch with our higher levels is that the predominant propaganda to which our eyes and ears are subjected is telling us that we don't have higher levels, that they don't exist. When we accept this, the connection with those levels is further reduced. What you don't think exists, you don't seek to reconnect with.

But things are changing fast. Positive and balancing energies have been coming into this frequency from other levels in ever greater power since the 1960s. It was they who awakened so many to another way of thinking which manifested itself as the phenomenon called 'flower power'. These energies were too much to handle for many. People became 'ungrounded', and so the ideals of the 1960s lacked a direction that would bring about fundamental change at that time. In this decade and beyond, it will be different. Those energies coming into this frequency to balance the negative domination have been affecting more and more people through the 1970s and 1980s in ways that gave us the Green movement, the animal welfare movement and the explosion in vegetarianism. Now as we move through this decade the Great Awakening of humanity is upon us, as the interaction of negative and positive energies quicken the vibrations around the earth.

Already, all over the world tens and tens of millions are beginning to think, question, and see life in a new way. Soon it will be hundreds of millions and more. There is a mass reconnection underway between lower and higher selves as the energies around the planet become less disruptive to that connection. The very fact you are reading this book means that you are awakening.

It is not even a case of having access to new information. What I have set out in my books and talks is merely to help people *remember* what they already know. This is why some people have their entire lives changed by one book or one talk. They have remembered what this imbalanced frequency and the illusions of the system have led them to forget: *who they really are*.

The speed of this awakening can be increased if people make the decision to go with the quickening vibrations and not resist them. There are many ways we can help ourselves to open up to a 'new' or, rather, re-emerging, understanding.

11

A Time to Feel

You will probably have come across the term 'going within'.

When someone not in the know asks "what can I do?", and they are told "Go within", they often find such advice confusing because they don't know what it means. Linked to 'going within' is the word 'meditation', which again can be a little daunting to people just turning their minds to these subjects. I prefer the description 'sitting quietly'. This is an effective way of 'going within' – re-connecting with those higher levels of ourselves that hold all the information we need to overcome the illusions of the physical world, and do what we are here to do.

It is when we sit quietly and let our minds and thoughts go wherever they wish, that we can hear, or most often, 'feel' our higher selves. I recommend it. It can be a wonderful feeling as you experience the energies around you and the love that enfolds you. The more you do this, the more your sensitivity will be heightened, and the more powerfully you will feel these energies and sense what your heart and your higher self is saying to you.

Amid the noise, clamour and self-imposed pressure of modern life, the ears and eyes will always dominate, at least at first. The time will come when you are so re-connected with your higher mind that even the distractions of noise, desires and pressures will not get in the way. But if you are just awakening, it is important that you take time out to sit quietly without any agenda and *just feel*. This is not an emotion as such, like anger, love or regret. It is more an intuitive feeling like we say 'a gut feeling'. What you think can be affected by all manner of confusions. The most important indicator of your intuition and the urgings of the higher self is what you *feel*. What you think and feel can be different. What you think is best for you might be what the system and your programming wants you to think is best for you. What you *feel* is best – that feeling and urging from your heart and intuition will come from higher levels. They are the levels that really know what's best for your evolution, even though that might conflict with what your eyes and ears are telling you.

'Rational' thought is not the only kind of thought worth the name. It is still 'rational' to think that increasing production and consumption is the only way to survive, when actually it is the quickest way to global oblivion. Such an idea is, in fact, the very definition of *irrational*. How can we take lectures on what is 'rational' from a system that is dismantling the world? Suspend your judgment on what is or isn't 'rational', and allow what you feel to come to the surface, no matter if it seems incredible in the face of what is considered sane by the system's insane standards.

If you have to make a choice between what you think and feel, follow what you *feel* every time. The moment will come, as the re-connection into wholeness gathers pace, when what you think and feel will always be the same, because at that point the whole wisdom of the mind will be channelled through the physical level. There will be no divisions. In the same way your higher mind will be seeking the most effective connection with the Infinite Mind of Creation. When your physical level re-connects with your higher mind, and your higher mind finds a powerful connection with the Infinite Mind, so much will be possible to understand and achieve that is beyond our comprehension today. That prospect is there for you in this lifetime if you want it enough.

People have written to me describing experiences while sitting quietly or even when walking down the street, which have shown them the truth that we are all One. A man wrote to describe how he was walking along when suddenly he felt part of everything. He was still the same consciousness, still the 'person' he had always been, but at that moment, he said, he was also the pavement, the houses, the cars, and the people walking past him. There was no division between anything. He was part of a seamless stream of energy.

Sitting quietly and just 'being', without distractions, is a way we can raise ourselves to re-connect. There are many meditation groups (sitting quietly groups!) which you may like to join or, like me, you may prefer to work on this alone. It is up to each person to decide what feels right for them.

Some take meditation very seriously, and it becomes everything. That is their choice and it may well be right for them. But sitting quietly is part of the process of awakening, that's all. If we concentrate too much on one twig we may not see that the twig is part of a forest that is waiting to be explored. To be honest, I have done very little sitting quietly compared with some, but it has been valuable when I have. Again, follow your instincts. They will guide you to do whatever you need to do at this time.

Follow those instincts and don't deny them, and all will be well.

12

Getting Together

There is an ever-increasing array of groups and organisations aimed at expanding understanding and supporting each other, as the speed of the journey gets ever faster.

What I have said about 'isms' does not mean I am against joining or working with such groups. I think they are essential so that people of like mind can find support and information, in a world that is still very sceptical, even hostile, to such thinking. Getting together in groups is very advisable, and I have listed a few addresses at the back of the book for those who wish to contact some of them. You might even decide to start your own group.

The point I am making is that we need continually to follow our own instincts and intuition, and while it is helpful to work with groups of people and listen to what they say, those instincts must be the guide that decide our course of action, even if that means going against what every other person in the group says and believes. Getting together with others is a way of supporting and stimulating the interplay of feelings and ideas, but we ourselves must make the final decision, and have the confidence to go against the majority view, if we are to stop the group thinking for us and becoming a prison of the mind.

Be careful also about feeling inferior in the presence of people whose awakening began long before yours. I have heard many times what I call the 'I've been doing this for twenty years' syndrome. This is the idea that if someone has been working in these areas for a comparatively long time, they must be further along the road of understanding and more in touch with their higher mind. That may be so, but just as easily it may not. Under that criterion every thirty year old sportsman must be more talented than every twenty-one year old. This is clearly not the case. Does every seventy-five year old have a better understanding of life than every forty year old? No. The time spent working at something is only useful if that time is used well. There are people whose experience you can learn from, but time-serving in a subject is not, by itself, proof of true understanding.

Also, the time of awakening is no guide to someone's evolution. It is likely that the period of awakening was timed to fit in with a life-plan. I recall someone at a meeting telling me I was not as evolved as he because he had opened up to this subject years before. But that is irrelevant. How evolved we are or are not is the result of endless physical lifetimes and periods on other frequencies going back to the moment we first became conscious. There are minds working through the bodies of five month old babies today that are far more evolved than those working through physical forms that have been on the planet for eighty years. The age of the body in this life has nothing to do with a mind's eternal evolution. It is worth pointing out, too, that to go around saying you are more evolved than someone else, to see yourself and others in those terms, betrays anything but an evolved understanding.

So I would say listen to others and their experience without being intimidated by them.

This brings us to channellers. We all have the ability to channel, which merely means to tune your consciousness to another frequency and allow thought-energy to be passed through your consciousness to this level. Here your lower self, through your brain, decodes the thought-energy and turns it into either spoken words or what is called automatic writing. Some people have fine-tuned this ability and can do it more effectively than others, but not all those who call themselves 'channellers' are gifted, just as not all footballers are Pele. For every Pele or George Best in the world of football, there are millions who play for their local pub team on the park.

Being less gifted than Pele at football is not a danger to others. If you claimed to be a Pele it would take only a few minutes with a ball to see very clearly that you were not. If only the same could be said of channelling! If I had to write the word *beware* only once in this book, it would be alongside channelling. As you seek and awaken, you will come across channellers and channelled information about yourself and the planet. That's fine and desirable if you are selective.

The purest and most brilliant channellers are relatively few. To reach that stage you need to have re-connected very closely with your higher self, and through that to the Infinite Mind. The number of those who have achieved that is infinitesimal when contrasted with the number of people around today claiming to be channellers. To have reconnected to such an extent means that the information can come through the lower self (by now at one with the whole self) with a purity and clarity that is not muddied by the ego and the emotions affected by that ego.

That is not to say that people who channel before they have reached this state of being are to be written off. But the words they speak or write are likely to be, at least to some extent, a less-than-perfect replica of the original thought-energy that entered them at a higher level. I have found people who are far too ready to believe and act upon information channelled by someone. Channellers can tend to be almost worshipped sometimes and seen as fountains of knowledge, when this is often not the case at all. Setting up as a channeller can also be used, even unknowingly, as a means of manipulation. The information they speak or write is what *they* wish to happen, not necessarily what the other level is sending. They can wreak havoc in your life if you are not careful.

We would do well not to underestimate the determination of misguided, imbalanced, aspects of the One Consciousness to disrupt and destroy the awakening and transformation of humanity and planet earth, by using inexperienced or less-gifted channellers to pass on utter nonsense. There are an infinite number of frequencies in Creation, and not all of them are at the stage we would call *wisdom*. As someone said: "Death is no cure for ignorance". To what level of consciousness is the channeller tuning? If it is to quite a low level, they *might* be telling you a load of disruptive, mind-scrambling, life-scrambling nonsense. One thing to avoid at all times is devices like Ouija boards, which can be vehicles for, shall we say, less enlightened levels.

Let your feelings tell you what is right for you – be street-wise and selective. One good way is to listen to a number of channellers without telling them what anyone else has told you. It is the same with tarot card readers and those who use similar methods of communication. See what common themes emerge and look at those rather than the detail, which can quite easily be distorted as the thought-energy passes through the levels of being.

The good news is that as the awakening proceeds more people will be re-connecting with their higher levels, and the purity of channelling will increase. But really channelling, in the sense of moving into a semi-conscious trance, is only one way of bringing down information to this level, and I feel this will gradually fade in the future. When you get closer to your higher self in vibratory terms and become at one with it, you will find that you no longer need to go to channellers except for perhaps confirmation of what you feel. When you are 'in sync', information is passed down to you all the time.

As I write this book I don't go into a trance every time I sit at the typewriter, but it is still being 'channelled' from a higher level of myself. Sometimes as I sit here the words pour through my mind at such a speed that I am hardly able to keep up with them on the keyboard. This entire book has been completed in less than a week. But it was not done in a channelling 'mode'. I am sitting here feeling quite normal, and stopping every few paragraphs to chat about the events of the day with Linda and sip a nice cup of tea! This is the way humanity will be communicating as the awakening goes on, I feel. There will be no higher and lower self – only self, a self with a direct connection with the Infinite Mind. This, I believe, is the real meaning of the constant theme through the centuries in religions and legend, of 'returning to the father'. It means re-linking with the consciousness of all that is, and so tapping into quite unbelievable knowledge, wisdom and potential.

What a great time to be here!

13

Lights in the Sky

There are many obsessions within the subject we are discussing that can divert us from continuing to seek, and so find.

We can get hooked on channelling one aspect of consciousness all our lives instead of raising our vibrations to tune to ever-higher levels of consciousness. We can become so obsessed with meditation that we do that at the expense of making practical use of what we feel and understand at these moments. We can also get obsessed with one part of the transformation, like crop circles, and see them as 'it', rather than one piece in the puzzle. It is important that we continue to see each area of work and research as a component in a vast jigsaw and not as the entire picture.

If we do this, each component part we come across can be a stepping stone to the next, and so on until the whole picture begins to emerge. This is happening to many people who became interested in a new way of thinking by trying to explain the crop circle phenomena. Their research has led

them on to an interest in UFOs and lights in the sky because of the sightings of such things in areas where crop circles and patterns have appeared. This, in turn, is leading people onto other areas of research – into areas that reveal that some enormous change in planetary consciousness is underway.

We all come into this way of thinking at different points. With me it was meeting a medium, with others it is crop circles, with still others it is alternative healing or UFOs. From that 'in' point, however, it is necessary to spread out and look at all aspects if we are to gain a balanced and whole view of what is happening and who we are.

Until recently, I have not felt the need to look closely at the UFO phenomena. I was spending all my time working in other areas and doing other things. Then suddenly information was coming at me from all angles about extra-terrestrials, space people from other worlds. You will probably find this in your own awakening. Apparently out of the blue, information about a particular subject will reveal itself through many different sources in a short time. It is clear that someone is trying to tell you something: the moment has come to encompass that knowledge into your overall view.

The more I have opened my mind to the subject of UFOs and extra-terrestrials, the more I am convinced that this is a key component to understanding the nature of the current transformation of the planet and to appreciating what has been going on in the other realms, frequencies, in the struggle between light and dark, harmony and disharmony. This spiritual confrontation is now being worked out on this physical level today.

My feeling is that there has been a fantastic struggle on other frequencies between the forces of harmony and disharmony which has been portrayed symbolically in films like 'Star Wars' and 'The Empire Strikes Back'. I believe most strongly that films and books of this kind, and others like 'ET' and 'Close Encounters of the Third Kind', are the result of knowledge seeping down into the conscious level of the writers. It appears as a science fiction story, but is in reality based on science fact – what they know on their higher levels has happened on other frequencies.

I believe that an aspect of consciousness decided that it would no longer follow the laws of Creation. These are not laws written down in a book, they are more like the laws of physics – the laws you must follow for harmony to reign in the sea of consciousness that is Creation. This misguided aspect who decided to challenge those laws we will call 'Lucifer'. When you go against the natural flows that bring harmony you of course create the opposite, disharmony. If you can then understand how to harness that disharmony and negative energy, how to feed off it and become stronger, you can begin to wreak havoc. You can so destabilise other areas of consciousness that they become part of your way of 'thinking' and add to your numbers.

There came a time, I feel, when those who wished to preserve harmony decided that this Luciferian Consciousness had to be challenged. This was the so called War in the Heavens talked of in *Revelations*. Planets were destroyed, either directly by the forces of disharmony, or they influenced others to do it. Look at what we have done to this planet, for instance, often in the belief that we were trying to improve things. Has humanity been guided by a force for good in what it has done to itself and the Earth? Hardly. The same and worse has happened on other planets to the point where they ceased to exist, sometimes exploding apart as a result of nuclear explosions. A great rift in the universe resulted and we are now seeing the process of restoring harmony once again.

The Earth has for a long time been possessed and controlled largely by the Luciferian Consciousness and human evolution has been diverted as a result. A call went out for volunteers on other levels who were willing to come to this planet and work for its restoration. This meant incarnating into a very hostile situation among those under the control of the Lucifer Consciousness and these volunteers were constantly challenged by that consciousness while in incarnation. The man we call Jesus was one of those who volunteered and there were many others. The necessary level of determination, courage, and understanding to do this job was, however, extremely high and the number of volunteers who proved capable of passing all the tests in many physical lifetimes to the point where they could be trusted with the responsibility of playing their role today was relatively small. Some say it was the 144,000 referred to in *Revelations*. I don't know about that, but I have no doubt there are not that many compared with the population of the planet today, which numbers almost six billion.

You can often identify these people – maybe in yourself – as those who have been, or are still going through, horrendously difficult experiences. They ask questions like: "Why am I going through this terrible experience when all I want to do is work for the good of the planet?" Recognise that line? These are the final tests, the final honing and strengthening of the spiritual steel that is going to be necessary in the rest of this physical lifetime as the power of the Luciferian forces is removed from this planet and the collective mind of humanity is released from the prison these forces have helped to create. It was they that destroyed Atlantis and brought the disharmony here that has helped to de-link humans from their higher levels.

Always when these volunteers have been in incarnation over the centuries they have been supported, often unknowingly, by what we call UFOs and extra-terrestrials. If you look at the Bible you will see many examples of people describing spaceships. I know the idea that Jesus was helped by spaceships may sound daft to many, because they see the comparitively primitive nature of life in Palestine at that time in stark contrast to the breathtaking technology that would be required for space travel. But we need to remember that what happens on Earth is only the point that it has reached in its own evolution. Everywhere is not the same. Beings from other civilisations that are far more advanced than us technologically, and from other more highly evolved frequencies, have visited the Earth and taken a close interest in its welfare ever since the Luciferian consciousness took it over.

Today, more than ever before, the Earth is surrounded on other vibrations by what we would call ETs, those from other planets, frequencies, and areas of this universe who are playing their part in the cleansing of this planet and the removal of the Lucifer influence. They and their craft have the ability to switch frequencies, come to this one, and then return. When this happens a light in the sky or a spaceship will apparently disappear when in fact it has merely switched to another frequency so it can no longer be seen on this one. UFOs can also be seen through our psychic sight on other levels, and on these occasions one person will see them while someone next to them will not. The sightings of UFOs are growing by the year and there is no doubt whatsoever in my mind that at the highest levels of military intelligence, particularly in the United States, they know of their existence and have had contact with them. Many people who have seen spaceships have been visited by the security services to be told to keep quiet because they know the likelihood of public alarm resulting from a realisation of what is obviously true – that we are not the only life-forms in Creation.

The time is approaching when these spaceships will land in the full view of everyone, but before that can happen human consciousness has to be prepared. This event would cause such a collosal explosion in the human belief system that it would create tremendous mental and emotional upheaval. This is why it is being done in stages, a sighting here, a few more sightings there, films about them that reach a massive audience, so that it dawns in the human consciousness as smoothly as possible that we are not alone and, at least most of these ships come in peace and love.

This is not the case with all of them, however, and I think it is right to emphasise that. Just as there are misguided people on Earth, so there are elsewhere. I recall channelling some information once in relation to UFOs which said: "Just because there are civilisations further ahead of you technologically does not mean they are any more wise in how they use that technology". I believe there are Luciferian-influenced visitations from some UFOs which do not have the best interests of this planet at heart. There are some horrible stories of abductions and the physical abuse of horses and other animals which I feel are the work of such beings. These tales have stimulated a lot of fear about UFOs which is unfortunate because it makes it harder for the majority to be accepted for what they are – beings that wish to love the Earth and humanity back to physical, mental and emotional health. I feel that the contact the military intelligence services have had with UFOs and their occupants are of the misguided technology-without-wisdom evolution. But their days of creating fear and adversely affecting the Earth are coming to an end because harmony and love are casting a shield of protection around this planet that will ensure that the time of any Luciferian manifestation will soon have passed.

If you are someone who has volunteered to play a part in the healing of the world then you will be in some form of contact telepathically or through direct channelling with so-called ETs. There is a sort of mission control which is co-ordinating this work and I know it is often referred to as Ashtar Command. What form it takes or what we perceive it to be is not important. Its effect is what matters. You will have been pre-programmed to awaken at the right time and you will be protected as much as is possible from anything that will seek to stop you doing the work you have come to do. You are being watched every second of your lives. Yes, even when you are doing that!

UFOs, ETs and the whole cosmic question are going to be coming more and more to the fore in the years ahead and we are going to realise that the very idea that we are alone in this universe is utterly ridiculous. What's more, if we are going to heal the planet, we are going to need the help of those the system desperately wants us to believe don't even exist.

14

Living in the Past

Another obsession I have found in some cases is that of past lives. It can be very helpful to know something of your past incarnations because it can give you a better appreciation of what you are doing in this life, and why you have been through certain experiences.

Some people, however, get hooked on wanting to know who and what they have been in the past. It becomes everything to them. I can understand this, especially when you first begin to awaken. I remember I wanted to know everyone I had been! But you soon realise it is not necessary to know and, except in the terms I have outlined, of no importance. At various times I have been told by different psychics that my consciousness, my mind, had incarnations as the Greek philosopher, Socrates, King Arthur, and that much misrepresented figure we call Jesus. That must make me about the 20,000th person in incarnation today who has been told he was Jesus!

But even if that information about my previous incarnations were correct, so what? They, like all incarnations, are periods of experience. Those who have incarnated as well known figures of history, have also had simple lives of all kinds in many civilisations, creeds, cultures and belief-systems. That is the only means to evolve in a balanced way. You may well have incarnated as a famous figure, but that could be no more or less important to you than another life which history did not record. To talk of 'us' being someone in the past is very simplistic, anyway. There is a little more to it than that, I think.

When I have spoken at public events, some sensitive people in the audience have seen figures around me. It happens often. On some occasions they will see North American Indians, on others a Roman Orator, and, when I am questioning religion, a religious figure or a nun! I feel these 'figures' are the manifestations of previous incarnations that come forward when I need to use what I learned in those lifetimes. Previous incarnations are like a bank or library of experience which we can then call on at any time. If Socrates was an incarnation of my consciousness, I have no doubt that I have been tuning into that part of my 'library' for some of the philosophy in this book. That doesn't make me 'special'. Everyone is the same. It is not what we have been or have done that matters most. It is what we are and are doing today. Now. This second.

Alongside research into relevant past incarnations comes astrology. This, too, can give us a basic idea of what we have come to do, and the area to work on in this life. The planets send out certain vibrations, thought-patterns, and so at any one moment, depending on where the planets are, some of those vibrations will be more powerful around the earth than others. At the second we are born we inherit the energy-pattern in the environment around us at that time. This birth-chart pattern will then interact with planetary movements through our life in a different way to someone born at another time, when they would have inherited a different energy-pattern at birth.

By the skilled reading of where the planets were at your birth, a good astrologer can give you a fair idea of the main outline of why you are here this time. This they can see by the energies you have chosen to absorb at birth, because those energies will have been selected by you to give you the best opportunity to complete your life-plan successfully. An astrologer can also advise you on when it is best to do certain things.

They will see when the energies most appropriate to a situation will be at their most powerful within and around you, as your energy-pattern is triggered by the moving planets. It is worth going to see an astrologer to add to your knowledge, but again, remember there are skilled astrologers and there are those who merely call themselves astrologers. Finding someone through personal recommendation is the best way.

15

The Long and Winding Road

One of the first communications I was given when I visited my first psychic was: "The spiritual way is tough, and no one makes it easy".

They were not kidding!

It is tempting to think, when you open yourself to this wonderful new vision and understanding of life, that everything will be easy from then on. Well it will and it won't – if that doesn't sound too daft. It depends at what level you are talking about. Life can get very much harder on some levels, particularly in the first periods of awakening.

You will be taking on board powerful positive energies – tuning in to higher levels of knowledge and wisdom, and re-synchronisation with your higher self is, by itself, an extremely positive experience. If you are to stay grounded and balanced and not float off in a spiritual dream, this positive energy needs to be balanced by negative. More than that you need the interplay of positive and negative energies for evolution to take place. Life will not become free of negative experience. What will happen is that negative experience will cease to affect you in the way it has in the past. True peace is not freedom from hassle – it is being unaffected by that hassle!

In the three years after my own conscious 'activation' in 1990 enormous amounts of quite incredible positive energy were pouring into my energy-field, but around me on the physical level I was going through some equally incredible negative experiences. One was there to balance the other, and the power of the positive and negative interaction launched me like some spiritual space-shuttle, to a whole new understanding. Parts of this experience were sheer joy. Others were horrible. Both were necessary.

The term 'go with the flow' is most appropriate. It is like riding a spiritual surf-board on a tide of energy taking you in a certain direction. It can be a bumpy ride, and you have to work hard to stay on your feet, but it is so much easier than trying to surf against the tide. Life on this level can be tough when you follow your higher self, because what you say and do will become increasingly out of step, at least initially, with the rest of society. You will be guided by a different set of principles and understandings.

I don't, however, wish to paint a picture that pain and anguish is a guaranteed part of your future. I can say from what I have been through myself that the moment comes when what would once have been felt as a negative, painful, situation can be coped with in a way you would never have thought possible. Also, as the planet is re-balanced, the scale of negative experience will begin to diminish anyway. The more you have the confidence to be true to yourself and leave behind your worries of what other people think of you, the fewer negative situations there are to affect you as you get closer to balance. Most pain and emotional trauma comes in the earlier stages of the awakening process, while you are still affected by the 'values' the status quo imposes on people, and people impose on each other. You will find that negative experiences will be taken in your stride as merely another experience on the road to enlightenment.

You may think many things that turn out to be less-than-accurate, but that does not necessarily mean you misinterpreted the information. Sometimes thought-forms will be sent to you from higher levels which, while not strictly true, lead you to act in a way that your life-plan demands. For instance, at a time when I knew little about a character called the Count St Germain, I saw a leaflet in France advertising a tourist attraction called Le Chateau de St Germain. Now there are many places in Normandy called St Germain, and I would not have taken much notice had I not had the impression that this chateau was the former home of Count St Germain. As a result I went along to see it. The chateau was not the home of the count, but merely named after the local village. The point was, however, that my misunderstanding led me there, and it turned out to be an important point on the earth's energy system, which I have visited several times since.

Symbolism is another essential form of communication you will begin to experience. You will see things in your everyday life that strike you as significant, but you won't know why. Symbolism is the universal language of Creation, and part of its value is to encourage you to expand your consciousness by trying to work out the meaning for you of the symbolism you see. If we were given all the information we needed on a piece of paper from on high, answering all our questions, it would have little effect on our vibratory state. It is the *seeking* which expands the mind. So if you keep seeing the same situation or object, or you see a sign or word that really hits you, try to open your mind to what its significance could be. What are they trying to tell you?

You are likely to think and do some strange and funny

things when you begin to open up. In another book, I have likened this to a bowl of calm water. When you turn on the tap, the calmness is disturbed until the tap is on full, and the water is thrashing around in chaos, turmoil and confusion. When the tap is turned off, the calmness returns, but now at a much higher level in the bowl. When your physical consciousness, the lower self, frees its blockages and opens itself to higher levels, you are a little like that symbolic bowl of water. Energies come into your energy-field, and it can be stirred up to cause temporary chaos, turmoil and confusion. You are switching levels of understanding, often with great speed, and your whole view of life, security, everything, crumbles before your eyes. This can be very traumatic for you and those around you. This psychology of change is something that needs urgently to be addressed and understood. So many are going to be affected by it as the months and years unfold. We are going to see relationships breaking up because one person is not tuning into the changes like the other. But we will also see relationships getting ever-stronger when both partners awaken together.

If you are going through this awakening now, or you know someone who is, the key word is *balance*. Keep thinking balance. Negative-positive and spiritual-physical balance. I think a good way to avoid great extremes at these times is to use what I describe as the 'backburner principle'. As the energies flow into your energy-field, they are going to be carrying large amounts of 'new' information, and you are not going to be able to process it all and put it into perspective. Some of these energies will carry the memories of past events on earth, especially those which had a big effect on the human mind. You might find yourself believing you were anything from Adolf Hitler to the Virgin Mary. You might be travelling all over the place because the pendulum you are using for communicating with other levels swings a certain way to answer 'yes' or 'no' to your questions. Yet later you realise that the pendulum is lethal in unskilled hands – it so easy to affect the swing with your own thought-patterns.

My advice, having been through this awakening in a massive and very public way, is to use the backburner principle. Take the information you are sensing, and any you might receive through channellers or other sources, and let it simmer away at the back of the stove before you leap into action. Appreciate that your energy-field is like that bowl of water, and many things will come into your consciousness that need to be allowed to simmer for a while, until they mix with other ingredients and information to give you something like a true picture.

This will help you to avoid some of the extremes, but I should emphasise that if experiencing those extremes is part of your life-plan, as it is with me, then nothing I or anyone else says to you will make any difference. It will happen. The more you re-connect with your higher mind, the more easily the higher mind can override the conscious level, to ensure that a life-plan experience takes place. This is why we get ourselves into situations and ask: "How did I get into this – why didn't I see this coming?" The answer is we were not meant to. Our conscious level is switched off, hypnotised, you might say, so that it cannot see what is approaching, and so cannot take action to avoid something it has chosen before incarnation. The lower self has the free will to resist this, but the higher levels know what we have chosen to experience, and they will do all they can to make that happen. This inner conflict between lower and higher can create serious physical or emotional problems sometimes.

I have talked at length about negative-positive balance, but equally important is the male-female, spiritual-physical balance. We have chosen to come into physical incarnation, and we do ourselves and others no favours if we then try to ignore that we have done so. Some people become so obsessed with the spiritual that they cease to be grounded,

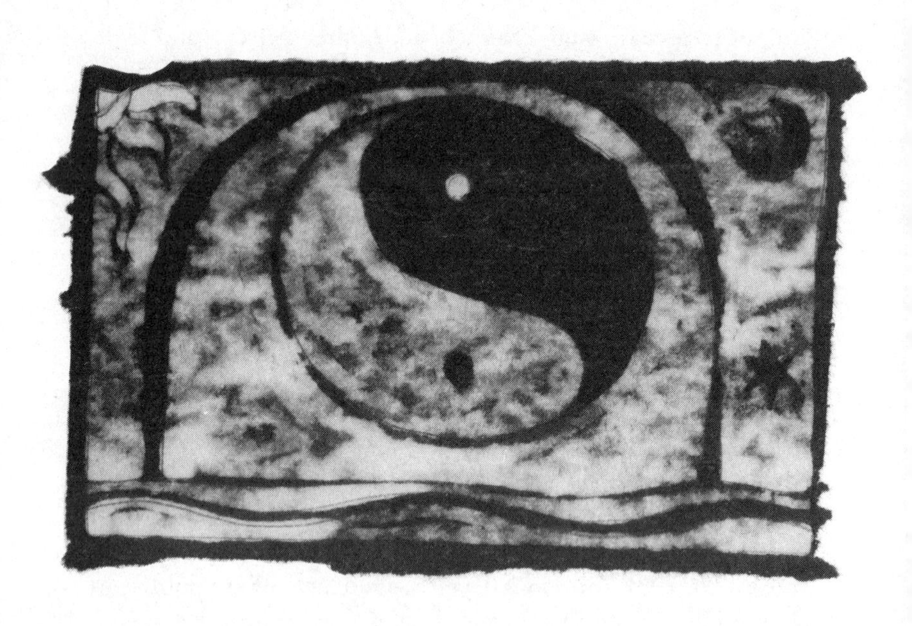

anchored, on the physical plane. Their consciousness spends as little time as possible on this level, which means their potential to contribute to what needs to be done in a practical way is very much reduced. This is just as imbalanced as being unable to see beyond the physical, and laughing at the very existence of the spiritual. It is balance we are after.

My advice is to find at least one major thing in your life that is very much on this physical level, so that it balances your spiritual expansion. With me it has been football. I take time to forget all about channelling and other levels and play, watch, or discuss football or any other sport. We need to live in this world as well as tune into higher levels. I have met scores of people who have only one topic of conversation – the spiritual, in whatever form. That is all they have in their lives, all they wish to talk about. There are those who say they never watch the television or the news, and they rarely mix with anyone who doesn't share their basic view. But how can we help others and relate what we say to the daily lives of others, if we know nothing about what is happening in the world outside our own belief system?

Balance, balance, balance, in all things. Nothing is black and white. Everything is made up of a range of colours.

The most effective way to maintain balance and perspective is to have a sense of humour. Laugh! Laugh at yourself and feel good about laughing at yourself, because you have that self-confidence. Some of the situations we get ourselves into in this business can be hilarious. I have known UFO watchers look into the sky and say a UFO is approaching, and trying to make contact, only for the UFO to be a hand-glider or helicopter! If we take ourselves too seriously, that is a lack of balance also.

One thing that comes over constantly from the higher levels is their sense of fun. I was standing at a sacred energy site once, after some amazing energy channelling. I was starting to feel very hungry. It had been an event of great power and reverence, but immediately it was over I was shown a vision of the buffet bar at Salisbury railway station, with a cup of tea and a doughnut! I took the hint, and the doughnut was superb.

A sense of humour protects you from disruption by negative influences. If you don't retain a positive outlook, the fear of 'dark forces' can take you over and become a dangerous obsession. Yes, there are misguided, highly imbalanced, aspects of consciousness trying to destroy the planet, and the universe with it. Their thought-patterns will try to disrupt you, especially in the awakening period. But if we fear them and become obsessed with them, we simply attract them. I have met people who thought everyone they came into contact with, who disagreed with their beliefs, was under the influence of 'the other side'. In fact, if anyone is under influence of negative thought-patterns, it is they because they see the world and other people only in negative terms. If you think that way, you surround yourself with those energies. I have seen this obsession destroy many lives and crush much potential for positive action. Fear of any kind affects the connection with the higher levels, and humour and laughter can prevent this.

The more you relax and go with the flow that is guiding you, the easier it all becomes. You no longer have to stand there anxiously with a pendulum asking what to do next. Your intuition is so well tuned to the flow of energies that you simply 'know'.

Laugh, and the world laughs with you. Our thought-patterns create our own reality. If we think life is to be endured, that is what it will feel like. But surely life is to be *enjoyed*. Think that, and it will be so.

16

Thinking as One

We have come a long way since the opening line of this book, and yet the ground we have covered has almost entirely been about self-healing, freeing ourselves from the programming, and opening up to a new vision.

The title may be *Heal The World*, but without all I have outlined so far, playing a part in healing the planet is not possible. Now we are at the stage where self-healing should be well underway, and we can look at how we can directly use this knowledge and thinking to help Mother Earth.

Every species has a collective mind, and humanity is no different. At some level the minds of what we call the human race – all those in incarnation – are connected. At an even higher level, all species are connected, and at the highest level everything is connected to everything else. This is seen, for example, through what is called the Hundredth Monkey Syndrome. It has been shown that once a certain number of a species learns to do something new, suddenly the rest of the species can do it without ever being shown. This is possible because the learning, the information, experienced by a certain number, is enough to affect the collective mind of that species and to become available to all of them. This is one of Creation's most important truths.

It is crucial in understanding how to heal the earth, and why the earth needs to be healed. The damage to the mind of the planet is caused by negative and disruptive thought patterns being generated by the collective mind of *humanity*. These negative patterns are being added to with every thought of every human being who thinks the way the system wants them to think.

The collective mind – 'public opinion', you might say – is dominated by the thought-patterns that illusions coming in through the eyes and ears have programmed 'individual' minds to create. These thought-patterns (energies) at the collective level have in effect declared war on the mind of the Earth Spirit. She has become so overwhelmed by this bombardment that her mental and emotional levels have been destabilised. As a result, the balance she brings to the physical planet and sends out to the universe and beyond has become destabilised.

We can see the effect in a dying planet, in the inability of the Earth to restore and repair in the way she once could, in growing extremes of weather and, as seems highly likely in this decade, in some mind-blowing geological activity. The physical Earth is not going to be allowed to die. It is the heart chakra, the balance point and source of the energy called love, for at least a significant area of this universe. Already the imbalances have been exported to a wider area of Creation, and for the Earth to go out of incarnation would have highly negative knock-on effects in this frequency and in at least those close to it. The question is not "Is the Earth going to survive?". She is. The real question is this: "How much of humanity will survive the changes necessary to protect the planet?".

That is our choice and ours alone.

As the energies around the earth increase in their vibrations there will be a sort of filtering effect. Those who stay at the level where their thought-patterns continue to add to the imbalances of the collective mind will find themselves falling out of synchronisation with the gathering vibrations around them. Only by changing and expanding the way they think will they be able to stay with the rising vibrations. There will be those who choose not to, and this will, in the short term, lead to more negative behaviour. But before too long the difference between their vibratory state and that of their environment will be so wide that their physical body will be unable to take it any longer. They will go out of incarnation.

This is not punishment because they are 'bad' people. It is that the Earth is moving to a higher vibratory state – frequency – and at this stage in their evolution they have been unable to go with her. Those minds will move to a non-physical frequency, and continue their journey at a level suited to them. No one can say what number this will involve. That is the decision that all people have to make for themselves. I stress that not all those who go out of incarnation will be 'out of sync'. Many will have chosen this time to depart even before they came.

Others who awaken will have the opportunity to expand their consciousness along with the Earth and go with her to the higher level she is destined to enjoy. But even then, it will require us to keep moving and seeking constantly, because this transformation is now going into higher and higher gears. It is speeding up with every month.

Everyone who awakens and begins to think positively and with a new awareness will be affecting the collective mind of the human race. So when I am asked by people "What can we do? What effect can one person have?", I reply "You can change the world". By changing your thinking you cease to be part of the problem, as you stop generating destructive thought-patterns and you become part of the solution by sending out healing thought-patterns. Every time another mind makes that change we are getting closer and closer to that point when the dominating pattern of the collective mind switches from judgment and destruction to balance and love. When that moment comes, the way that humanity thinks as a whole will change with a speed that will make the period of change in the former Soviet Union and the Eastern Bloc countries seem pedestrian.

That itself was an example of the collective mind of those countries being transformed. Once a certain number of people had realised that the so-called power of this communist empire was nothing more than an illusion, and that it was only there because the people allowed it be there, events happened very quickly. The switch of perception in collective consciousness brought people onto the streets, which then changed the perception even more, and so on. Eventually, in a period that would have been laughed at only weeks before, the apparently all-powerful Soviet empire came out with its hands up. The same is going to happen to the Western system of mind control on a global scale.

The effect of the information and thinking that the awakening is presenting to us is so powerful in the face of the illusions of the physical world, that it will not take fifty-one percent or even nearly that number to change their thought-patterns, for those illusions to be overturned within the collective mind. It will actually take comparatively few.

So another way of healing the planet is to pass on what you know and have experienced to others, and allow them to see that there are other views and explanations that the system has been denying them, for its own survival. Have the courage to speak out for what you believe, to anyone who wishes to listen. If some people turn away from you then that's their decision. Why should you be held back and controlled by what they think? We can now all have access to a printing press – the photocopier – which can network information in ever-expanding circles, like dropping a pebble into a pond. Pass information to others and ask each of them to print copies and pass them on to their friends, with the same request. Those sheets of information will also carry the thought-patterns that created them and this will affect people who are ready to awaken, too.

All I would say about this is, respect everyone's right to reject what you say. Forcing views upon people has no significant effect on raising their vibratory state. They have to make that decision themselves. Take every opportunity you can to speak or to write in the media, be it letters to local papers, taking part in phone-ins, whatever it may be, and more people will realise there is a whole new and glorious way of looking at life. Get involved. This is a time in opt in, not out. Start in your own community and take it from there. Every single one who changes their thinking in the light of what you do will speed the day when the 'big switch' occurs in the collective mind of the human race.

17

The Web of Life

Changing the way people think and feel is fundamental to earth healing, but it is not the only way.

If you know anything of acupuncture, you will be aware that the physical body has a series of energy-lines known as *meridians*. These go all around the body linking in with the chakra vortex-points to maintain the flow of life-force energies to the physical level. When these energy-lines are balanced and free-flowing the body is healthy. But when all is not well in the energy-system, the body is ill or diseased.

It is the same with the planet.

All over the earth on the unseen, sub-atomic level is a web of energy-lines and chakra points, and this system has been devastated by the lost understanding and destructive thought-patterns of humanity. Before Christianity, knowledge of this system was widespread. But the emerging Christian creed saw these energies and the acupuncture points along them as 'evil', because they were held sacred by people they believed to be heathens and savages.

Ironically, in an effort to suppress what they felt were evil energies, they built most of their early churches on important acupuncture and chakra points around the world. The legends and myths about St George defeating the dragon relate to the suppressing of these energies. The Chinese, I understand, still refer to this energy-system as dragon lines. I call them meridians, or as they are also widely known, ley-lines. The churches have indeed helped to weaken the flow of energies because of all the disruptive thoughts of fear and guilt that dominate so many religions, while those who have worshipped with love and joy in their hearts have strengthened the energies.

Some energy-sites are obvious because they are stone circles, like Stonehenge and Avebury, or mounds like Silbury Hill and high points like Glastonbury Tor. You also have natural energy-sites such as Uluru (Ayers Rock) in Australia, and rock formations in places like Arizona in the United States. They are all over the planet. Standing stones are like acupuncture needles balancing the flow of energy. Stone circles can be a sort of electrical circuit receiving energies into the system or boosting their power like an electrical coil.

There are many thousands of people, perhaps millions worldwide, being guided by higher levels to repair this system. You can join them. Use your free will to tell your guiding consciousness that you want to help, and if that is part of your life-plan things will begin to happen.

The system can be restored by balancing destructive thought-patterns which block the flow of energy along the lines. You can do this by sending out positive thoughts at these places, and these can be even more powerful if you visualise the negative blockage dispersing and the energy of love flowing again. Open up your heart and visualise the flow of love pouring from your heart chakra into the system, and the imbalances melting away. Don't underestimate what

your thoughts can achieve. As you become more sensitive you will feel the entire atmosphere around you change on these occasions.

These energy lines are, like everything else, lines of consciousness. Therefore they are affected by other consciousnesses. If they are bombarded by negative thought patterns their natural balance and flow will be affected because disharmony does that. On the other hand if you can go to these spots in reverence, love and respect, and send out thoughts of harmony, balance and love, then that will help to remove the imbalance and allow the energies to flow smoothly again. You can do this alone or with as many people as you like. If you hold hands in a circle to make a sort of electrical circuit, the collective power of your thoughts and energies will be far greater than the sum of the parts. It often helps to concentrate the thoughts if someone paints a word picture, a visualisation, for people to focus on during these occasions. It might be that they ask you to picture in your minds the site surrounded by dark energy and then ask you to visualise that energy dispersing and the whole area being bathed in bright, pulsating, golden light, with the energies flowing again. All this is happening on the non-physical levels, the levels of consciousness and thought. What you think in these moments is what you create, especially if you are a particularly powerful consciousness or many people do this together. You can change the whole nature of an energy site. In the same way there are those controlled by the Luciferian Consciousness who go to these sites to do exactly the opposite – generate disharmony. I don't know if you have ever been to a place where black magic and 'devil' worship has been going on, but my goodness you can feel the extreme negative energy thus generated. We can do the same for the Light to counteract that.

I say go to these sites in reverence and respect because they are a consciousness also. They are in awareness and recognise those who have respect for them and the laws of harmony, and those who do not. Nor is this a one-way process as I have said before. When we visit these places there is an energy exchange going on. We can help to heal their disharmony, but our interaction with their consciousness can activate many things within us that can help to awaken our memory and give us access to information held in the consciousness of these sacred places. Go to Glastonbury Tor and the other great sacred sites of the World and you can feel their knowledge and wisdom. If you are there for the right reasons, they will share that with you. It will reveal itself as the days, weeks, and months pass, as a knowingness, an understanding you cannot explain. You may not make the connection between that and your visit to a sacred place, but often that is where the knowledge was passed to you at a deep level from where it could begin its journey to becoming conscious. Sometimes that can happen immediately, while you are still at that place.

It is more effective if you are at a site, but you can even do this from home if you concentrate on a place, maybe with a picture of it to help your visualisation. I know it sounds incredible that you can restore energy-lines from home, but all you are doing is broadcasting a thought-pattern to that point to balance the one already there. When you turn on your radio it receives the radio broadcast signal, even though it might have been transmitted from another country. You are the same – a broadcasting transmitter with legs! There is no need to worry or think yourself inferior if you can't feel or sense these energies to begin with. When I started I went around with two people who stood there saying: 'Oh, feel these energies, they are really powerful'. I stood next to them thinking: 'I can't feel anything. Either it's me, or they are winding me up!'. Today I can sense and feel energies immediately.

Besides repairing energy-lines and removing blockages, there is a need to bring in energies from other frequencies.

The energies that are awakening the world must be passed through a physical form, or at least what is known as the etheric body, which is the next level up from the physical. This has to happen if the energies are to be changed into a form that the earth can absorb. It is like passing them through a transformer and changing one form of current into another.

The more people who awaken and raise their vibratory state, the more energy can come in. If they passed energy through you that was beyond your own vibratory rate, it could be a little like walking into a laser beam. It would do you and the earth no good at all. But if you are willing to undertake this work, energies can be passed through you at an ever more powerful vibration, as you yourself raise your frequency. There is often no need to even know this is happening, although as you become more sensitive you will be aware that something is going on. At its most powerful, it can be likened to being plugged into the mains.

In fact I have had some experiences in Peru, the United States, and Britain, that have been more like being plugged into a power station. In Stonehenge on the night of 26th/27th July 1993, I was in a circle of people holding hands in the way I have described to make an electrical circuit. They had all been drawn there by a belief that they needed to be in the Henge that night. Some had come from the United States. It was around midnight and suddenly I was hit by the most incredible energy. It was as if a massive bolt of electricity had struck me. My body arched backwards as I grunted and tried to stay upright. I could feel a surge of power building up inside me and as it came to the surface I found myself letting go what one of those who heard it called a 'primordial roar'. It came from the very depths of my being.

Another burst of energy just as powerful followed. My great friend, Yeva, who had organised the gathering that night, said she felt I should leave the circle and lie down on a rock that I always feel drawn to in Stonehenge. There, with

Yeva alongside, another five bolts of collosal energy hit me. I thought on two occasions that I was going out of incarnation, such was the power – but you are never given more than you can handle on these occasions, you will be relieved to hear! I had no idea exactly what was happening, but the next day we began to hear stories of power cuts in some areas not far from Stonehenge that night. Clocks stopped for an hour and people reported losing their late night television. Others said how they had seen a vortex of energy over Stonehenge at the relevant time, while still others a few miles away described seeing a bright light in the sky which suddenly exploded in all directions, lighting up the countryside as if it was daytime.

Perhaps the most astonishing part of that night involved a group of people in the United States who knew of the event being organised in Stonehenge. They had gone to a site in the States at exactly the same time all this was happening in the Henge. They too saw a vortex of energy spiralling above them. Then suddenly their silence was broken by a massive roar apparently coming from nowhere. They looked around to see what it could possibly be. They could find no explanation.

I tell this story because it is easy to wonder if it is all imagination when you go to energy sites. Am I really feeling anything? Do these energies really exist? Am I just kidding myself? I have heard these questions many times from those just moving into these areas of thought and I understand why they say that. They are fighting against indocrinated belief systems with apparently no physical 'proof' to support what they are doing. But these energies are real. The energy grid of the Earth is real. The need for as many people as possible to work with them is real. There is so much that humanity does not understand about the potential of harnessing these energies and the importance of their restoration to the future health and harmony of this planet.

We should remember also that this energy grid spans the world. Energy lines and points are everywhere. You don't always have to go to obvious sites like Stonehenge, Avebury, or Ayers Rock. There may be energy lines going through your home. Any place that you are drawn to is likely to be an energy point which has long been forgotten as myth has taken control of human understanding. I have channelled energies into the Earth in the centre of big cities many times. It doesn't half give passers by a laugh, I can tell you!

If your conscious level is willing to do this work, your intuition will begin to guide you, wherever you are in the world. As we repair the system and bring in the higher energies the transformation of the Earth and humanity will gather pace very quickly. There are certain times when planetary sequences are most beneficial, and energy-channelling is coordinated all over the planet at the same time. This then brings big changes as those energies take effect.

The most significant energy is the one known by many as the Christ. This is not a person, but an energy. It is pure love. Some say love is positive, some say a balance of negative and positive. I believe that the energy called love and its highest expression, the Christ, is all of them. It is whatever it needs to be. If a situation needs negative energy to achieve balance, then love and the Christ will be negative. If balance is required, it will be balanced. If positive energy is needed, love will be positive. It is whatever a mind, planet, frequency or situation requires. It is the great transformer of understanding, the balancer. The Christ energy is the ultimate manifestation of unconditional love on this planet.

The one we call Jesus could channel this energy, and perhaps it was from this point that it was given the name 'the Christ', as in Jesus Christ. But that mind is not the only one who can channel this energy.

It is within the potential of at least a large number in incarnation today. But they need to reach a certain vibratory state before that can happen, otherwise they will be fried by its power. I believe this was the energy that zapped through me in Stonehenge that night.

The last major inflow of the Christ energy was around two thousand years ago in Israel, because that was the most appropriate place for it to come in at that time. It entered an energy-pattern in the area known as Jerusalem. Today its main place of entry into the world is a different energy-pattern. The 'New Jerusalem' is England. The most significant area is a triangle of land linking Glastonbury Tor and points in Warwickshire and Hampshire. This is the heart chakra of the planet, from where love will pour in and out and change this world forever. It is no 'accident' that so many ancient sites are enclosed by this triangle, or that this is the crop-symbol capital of the world, and I feel this pattern connects with another of great importance in Normandy, going across to Paris. But the Christ and your own love can be channelled into the system anywhere. England is the main area, I feel, but not the only one. The world is covered in energy sites. North America alone has a network of immensely powerful sacred places.

Every day we can help to heal the Earth by taking a few minutes to sit down and use our consciousness to visualise her being healed, the darkness giving way to Light and love. You can do it at home, on the train, in the park, anywhere. The more people who do it with you the better. You can join groups like Fountain International who seek to get people together for Earth healing work – and I've put their address in the back of this book. Healing the divisions and pain of humanity is the same process, be it caused by war, famine, or prejudice. Visualise the situation changing to a positive outcome and that pattern will be projected to that area to check the imbalances that are causing the problem. The more who choose to do this, of course, the more effective it will be.

Think love, send out love, and you will be healing the world.

18

Over to You

You might have noticed that I have said at several points 'don't do this', 'avoid that' and variations on this theme. That is merely to overcome the need for endless repetitions of 'In my experience...' and 'My advice is...'.

For what you have read is not there to be followed unquestioningly. What you have read is what I feel is right. If you don't feel good about some of it, or even all of it, fine. In the end I don't know what is best for you – only you know that. But I hope that it will help people to avoid problems which it is not necessary for them to experience. This book is a classic confirmation that what appear at the time to be highly negative experiences do have a positive effect. I can only write in the terms I have set out before you, because of what I have been guided through since 1990.

Have I experienced some of the 'pitfalls' I have highlighted? No. I have experienced *most* of them!

For the first three years after my awakening became

conscious, if I wasn't climbing out of one spiritual trap, I was in the process of falling into the next. All this was preplanned, so that I could point them out to others on the basis of my own experience. This is far more powerful than getting things second-hand from a book or a 'teacher'. As a communication said when I was in the depths of despair:

"How can you help others if you have not been through it yourself?"

We are in a period of human and planetary change, in which the Earth Spirit and all forms of life will need all the help, love and support they can get. But most of all, they will need the confidence and courage to listen to that quiet voice called intuition, and follow what it is saying, no matter what their conscious thoughts may be thinking. It is worth repeating: *if it is a choice between thought and feeling – feeling is the one to follow*. That comes from the heart, that source of love, and from the higher levels of your own consciousness.

It is that intuition that will guide you to where you need to be, and what you need to do at this point of fantastic transformation. Everything is going to change – eventually even the nature of the physical body. Amid the chaos as the old order crumbles, you, too, will feel lost and sometimes fearful and confused. Others around you will feel the same. During these moments listen to your heart, and let it guide you. If you have a life-plan that involves you going through this transformation and helping to build the new and glorious tomorrow, then that is what will happen – if you follow your instincts and refuse to allow others to divert you.

The future is not pre-ordained. When people 'see' the future, they are, I believe, tuning into the energies, thought-patterns, that hold the *projected* future. It is a future that will be, only if humanity does not change in its thinking now. We are creating the future with every thought. If we as a collective mind cease to batter the Earth Spirit with negative thought-patterns, or reduce the rate at which we are doing

that, then she will more quickly regain her mental and emotional balance. This will result in a more comfortable transformation. The geological activity will be less catastrophic, changes in the weather will be less extreme, and there will be less conflict, war and negative behaviour of all kinds. If we continue on the path humanity has embarked upon, and remain largely under the influence of the status quo, the scale of all of those things will be simply awesome.

I feel that the awakening is already having an effect on reducing the impact of such events, but as things stand there will still be upheavals on all levels that will defy our imagination. It is in our hands, or rather our hearts, to reduce them still further. What I have said in this book is, I feel, the basis on which everyone can play a part in that.

On a practical level, everyone can reduce as much as possible the impact they are having on the Earth. The Green movement has many publications that can advise you on how to live in a way that is least harmful to the planet. It will not be environmentally friendly at this time – it will be less environmentally cruel. Ahead of us, however, is the knowledge of how to *create* through thought, in a way that allows total harmony between humans and the earth.

We are going to realise as the frequency rises and higher levels of consciousness become available to us that what we see as state-of-the-art today is really primitive. Take energy for warmth and power. All we are doing is taking vast areas of the physical earth and releasing the energy that matter contains very inefficiently so that most of the energy is lost in the process of releasing it! More than that we turn much of it into pollution and that does yet further damage to a planet which has already suffered by the taking of the matter in the first place. State-of-the-art?? As the frequency rises we will understand more clearly how to use energy in its non-physical state for all that we need for a decent life. We will no longer have to plunder the Earth to 'progress.' Indeed the knowledge of how to harness non-physical energy for free power without pollution already exists. But the consequences for the fossil fuel-dependent status quo are so fundamental that the knowledge is suppressed. Not for much longer, however. We are not, as some believe, going back to the Stone Age as the old order crumbles. We are going forward to a wonderful new tomorrow. But for now we can only do the best we can with our current knowledge to reduce the damage we do to the physical planet.

The most effective way to be less harmful to the earth is to simplify your life. It is also best for you. If you remove mental and emotional clutter, you have more space to feel and follow your intuition. And again there is a practical reason for simplification of our daily lives. Part of the speeding-up of vibrations is the quickening of what we call time. When people say there is not enough time in the day anymore, they are right. Time is getting faster. We won't see the hands of a clock moving quicker, because that is only a human measurement of time. What you might call 'cosmic time' is nothing like human time. It is cosmic time which is moving faster, and it shows itself in more and more events in a shorter and shorter period, and in the way we are sensing that the days, weeks, months and years are passing quicker and quicker. If we don't remove the irrelevant clutter from our lives, and instead try to continue to do as much as before, we will burn out. It will be like standing on a merry-go-round as it starts to spin. The world that once passed us by quite slowly will eventually be passing so fast it will be a blur. Simplification of our lives is vital.

I hope you feel you have benefited from my experiences. It makes them so much more worthwhile if you have. What I have learned, above all, is to let go of fear and guilt and seek to enjoy every moment. There will be times ahead in which your emotions will be anything but joyful. This will be only temporary, because we are being offered the opportunity to help to heal this beautiful and breathtaking planet, and help everyone to lift their understanding beyond the clutches of thought-control, and re-connect with the indescribable light and love that awaits us all.

Don't worry. Be free from fear. All is well. We are going to do it. This is our destiny and we are going to grasp that destiny. We are the generations that have been given this gift. We are the generations that will heal the earth and allow her to heal us and the universe.

We are all One Consciousness. I am you and you are me. We are all each other. Healing ourselves *is* healing the world.

Useful Addresses

Fountain International: PO Box 52, Torquay, Devon TQ2 8PE
College of Psychic Studies: 16 Queensberry Pl, London SW7 2EB
Friends of the Earth: 26–28 Underwood St, London N1 7JQ
Greenpeace UK: Canonbury Villas, London N1 2PN
National Federation of Spiritual Healers: Old Manor Farm Studios, Church St, Sunbury-on-Thames, Middlesex
Please send a stamped addressed envelope with all correspondence that requires a reply.

I have received thousands of letters from all over the World and I read every one. The volume of mail is such that I cannot reply to more than a few, but thank you to everyone who writes. I find them a great source of information, love and support.